FOLLOW IN THE
FUTURE
WITH THE
SHADOW

GEORGIY SERGEYEVICH GARBUZ

Published 2024
Printed in the United States of America

First Edition
ISBN (softcover): 978-1-963380-05-7
ISBN (e-book): 978-1-963380-06-4

For information, address:

Holzer Books LLC
8 The Green, Ste. A
Dover, Delaware 19901 USA

For information about special discounts available for bulk purchases, sales promotions, and educational needs, contact:
info@holzerbooksllc.com
+1 (888) 901-7776

PREFACE

Today, I am delighted to introduce you to "*Follow in the Future with the Shadow.*" This book narrates the real-life journey of a first-generation immigrant who arrived in a new country with little more than a dream. Like many immigrants in America, he began with no financial resources and struggled to overcome the language barrier. However, he seized the opportunities this new land offered and nurtured his talents, pursuing his passions and bringing joy to his life.

This individual's story encompasses the creation of projects, advancements in science and technology, medical innovations, and numerous other contributions aimed at improving his new homeland and the global economy. His journey is a testament to the potential that can be unlocked when one is determined and committed to their goals. He holds a unique international identity, with American and Russian citizenship, mixed Moldavian and Ukrainian heritage, and a background that includes being born in the former USSR and living in Kazakhstan before settling in the United States. His love for his adopted country is profound, yet he remains connected to his roots in the former Soviet countries.

His experiences have led him to champion various initiatives, with a special focus on the welfare of foster care children and disadvantaged youth worldwide. Having witnessed life in both prosperous and struggling economies, he understands the needs of people in different parts of the world. This understanding inspired him to initiate a groundbreaking space project aimed at enhancing global economic prospects.

This visionary firmly believes that all nations must collaborate in exploring and colonizing space, as well as providing quality education for every child to secure a brighter future. Although he is yet to have children of his own, he is driven by a deep love for the world's youth, particularly those in foster care, who yearn for a chance at a better life. He considers them his own, even without formal adoption, and he is determined to provide them with opportunities.

His grand vision is to grant foster care children from around the world the chance to study in International Space School and afterwards, International Space Agencies. He envisions a future where these children become highly valued contributors to their respective nations. His conviction is that a prosperous country starts with investing in education and ensuring that talented children have access to quality education, regardless of their financial circumstances.

Throughout his life, this man has operated in the shadows, working tirelessly to create a better future for people everywhere. However, the reality of the world often dictates that even with good intentions, some individuals resort to criminal activities when dealing with financial transactions. He has encountered both honest and unscrupulous people during his journey, especially as he embarks on a new political movement, "Za Detey - For Kids," which has a global focus. The movement's supporters share his vision of a brighter future and stand by him despite legal challenges in various countries.

This visionary is committed to creating projects, developing technologies, advancing medical treatments, and many other innovations that have the potential to generate trillions of dollars in annual revenue for every nation. However, he implores governments to reinvest the taxes generated from these projects into grants for children's education. He understands that this investment is not just about profits; it's about securing a better future.

By reading this book, you will delve into a world of secret operations, innovative business ideas, scientific breakthroughs, and technological advancements that promise to bring excitement and adrenaline to your life. The author has successfully planted his ideas in the minds of people across the globe, and as you read, you will gain a deeper understanding of the events of the past sixteen years.

This book will provoke thought and inspire you to contribute to your country's development. When you purchase this book, you are not only enriching your own knowledge but also supporting the author's endeavors, particularly in helping foster care children and disadvantaged youth worldwide.

Thank you for investing your time and resources in reading this book and for your invaluable support in making the future brighter for children around the world.

INTRODUCTION

L ife always presents numerous choices, and you have the power to decide how you will spend it. Making the right choices is crucial, as they shape the course of your entire life. This holds particularly true for immigrants who come to America seeking a fresh start—a new beginning that sets the foundation for the rest of their lives.

Life is a journey filled with ups and downs, reaching high into the sky. However, with a solid plan and dedicated hard work, you can turn your dreams into reality and become a part of a new reality.

As a first-generation immigrant, my journey has been a mix of rewarding achievements and significant challenges. Starting without proficiency in the English language, I built a solid foundation surrounded by friends and acquaintances. However, the path also led me through fifteen years of battling depression and financial struggles, living with limited resources. Despite having a vision for a global project, it turned out to be a mere replica. Pursuing this vision for a global project entail facing scrutiny from government agencies, which adds a layer of complexity and risk not, just for myself but for everyone involved.

Around the world, people engage in entrepreneurial projects with the hope of getting rich and making millions of dollars, often crossing legal boundaries for various reasons, only to later become subjects of investigation. Many may not comprehend that the meticulously prepared project plans already anticipate the repercussions. This phenomenon is rooted in psychology, and only professional scientists or those adept at strategic thinking can truly understand and predict the unfolding events.

As I often say, when you show people a path to potential wealth, they are prone to breaking the law, even if they have only a limited understanding. This tendency is so strong that individuals might find themselves entangled with legal statutes, such as the infamous "Statute 58" from the Soviet Union era. Government agencies are ever vigilant, with their eyes open, aware of your actions and words. They merely await a reason, a misstep on your part, to launch an investigation.

These agencies are known to exploit individuals, but the dynamics depend on how you utilize them and execute your plans. It's a delicate balance that could determine your safety. Reflecting on the global scenario over the past sixteen years since 2007, one can observe a pattern where people break the law for diverse reasons — some driven by the desire to become millionaires, others fall prey as victims of manipulation, and some merely wanting to be part of a significant, high-stakes game under constant scrutiny.

Individuals often embark on similar projects with the goal of accumulating wealth and achieving financial prosperity. To ensure success, they may seek the support and oversight of government agencies and surveillance. Ironically, many are unaware of the true nature and existence of my plan for an extensive *International Intergalactical Space Federation Agency (IISFA)*. Despite this, they find themselves accused of breaking the law, often falling under the shadow of the infamous Statute 58. The fortunate few, possessing substantial resources, may find themselves compelled to immigrate, while millions of less fortunate individuals, lacking financial means, face arrest and imprisonment—often unknown and unrecognized by the public.

The plight of these individuals, who lack connections and visibility, leads to suffering and a lack of understanding. It's crucial to recognize that all these outcomes were not only expected but also predicted. In every challenging situation, there is a solution, and it is possible to guide people who have encountered Statute 58 worldwide back to a normal life.

To address these global challenges, every country should acknowledge and embrace the authority of the *International Intergalactical Space Federation* (IISF). This organization, along with the IISFA, must be founded and meticulously planned under the leadership of the United States, Russia, England, China, Germany, and India—established according to the vision and guidance provided by me. This framework is designed to facilitate the reintegration of those who have faced Statute 58 accusations, ensuring their return to a stable and lawful existence.

The Allies, aiming to bring peace and prosperity to the world, envision a global order where each country is accountable for its territory. The president, as the head of the agency assigned to the territory, operates under the International Intergalactic Space Federation Agency. This framework ensures the safety of the people and fosters national prosperity. Grounded in the principles of *"Lyubov"* — love for the country, its people, and the world — this approach paves the way for a prosperous future. It extends beyond borders, with plans to expand territories through research and colonization of planets such as Mars and the Moon throughout the universe.

To turn this ambitious plan into reality, professional individuals from every country are invited to Minnesota. These professionals, familiar with real-world dynamics and prices in their respective countries, are crucial for the project's development. The goal is to build *National Space School*

Academies, National Space Academies, and establish an International Space government and agency.

The project emphasizes education, healthcare, and benefitting regular citizens, offering free education and healthcare. It calls for people passionate about science, eager to contribute to noble research, and interested in creating international intergalactic Space corporations. The focus is on those motivated not solely by financial gain, but by a genuine love for the world and a desire to make it better.

Money is seen as a crucial tool to realize dreams and improve the world, but the project's essence goes beyond financial considerations. The aim is to enlist individuals who care about people, the project, and the greater good. These individuals are envisioned to become historical figures, the first generals, and founders of the Space international government, leaving an indelible mark on the world. If you share this vision and commitment, I invite you to join me on this transformative journey.

My narrative unfolds uniquely in America, having been born in the Soviet Union, specifically Kazakhstan, and residing there until the dissolution. The stark differences between these countries manifest across various aspects—economy, individual rights, employment, and overall lifestyle. It serves as a testament to how people shape and adapt to the systems within their country, influencing laws, living conditions, and moral standards. Moreover, the story delves into the necessity for newcomers and immigrants to acclimate to the rules and norms of their adopted country.

The American dream is often romanticized—a desire to live within a legal system, experience societal acceptance, and join the millions who aspire to make America their home. It's perceived as an ideal place for raising families and children, where hard work is well-rewarded. In this dream, even those without formal education can secure good-paying jobs, presenting the prospect of a happy life. However, the contrast between living in different countries is stark. If Americans were placed in environments like Kazakhstan or Russia, they might struggle to survive due to the vastly different economic conditions, where individuals work for meager monthly wages, attempting to sustain their families amidst corruption.

Upon arriving in America, I found myself as a newcomer, starting my life anew with a mere eighty cents in my pocket as I stepped off the aircraft at the airport. The challenge was clear—I needed to build and create something that would not only bring financial stability to my life but also contribute to the well-being of others or my partners. It was imperative for me to learn and understand the system, connect with people, forge friendships and partnerships, and embark on the journey of working for a living while comprehending the intricacies of my surroundings. This journey, from humble beginnings to establishing oneself in a new land, encapsulates the essence of adaptation, resilience, and the pursuit of the American dream.

My aunt, Valentina, aided our family's migration, despite being a newcomer immigrant herself working hard to sustain her family. Her family has already done so much for us by inviting us to live in America as legal refugees, and we are truly grateful. Unfortunately, she is facing serious health challenges, battling brain cancer, and we need to assist her in obtaining treatment. The medical expenses are exorbitant, and her healthcare coverage doesn't include the necessary treatment.

Despite her illness, she has been working hard and saving money to undergo surgery in Moldavia, where the medical costs are significantly lower compared to the United States. For instance, a procedure that costs a hundred thousand dollars in America only amounts to five thousand dollars in Moldavia. Her situation is a stark reminder that illness knows no bounds; it afflicts people without warning. In reality, every human is dealing with something, whether it's a manageable condition or a severe disease lacking a definitive medical treatment.

The journey continues in a new country, among new people and a new nation. Acquiring citizenship marks the beginning of a new life, requiring finding employment. The language barrier adds to the challenge, but in Minnesota, we discovered a large Russian community where everyone knew each other. This familiarity facilitated job hunting, and despite my limited English proficiency, I secured a cleaning job for eight dollars an hour—a humble beginning to sustain myself.

Scientists are continuously researching and developing medicines and treatment methods. The enormity of tasks in the medical field is evident, and my journey begins in a new country, with new people, where I must assimilate and become a part of the nation after obtaining citizenship.

My life starts afresh, and the challenge is finding a good job, compounded by the fact that I don't know English. As we reside in my aunt's house and attend church, we begin meeting new people. Surprisingly, there turns out to be a sizable Russian community in Minnesota where everyone knows each other, making communication easier despite the language barrier.

Thinking on my feet, I secure a job in cleaning for eight dollars an hour. It's a modest start, but I need to earn a living. I draw upon the knowledge I've gained throughout my life—from my father Sergey, my mother Lyubov, my father's friends, and from reading books. Despite being underage, I've absorbed a wealth of information from my father, who exposed me to various aspects of life by taking me to meetings with mayors, politicians, KGB and KNB generals, police chiefs, directors, and even drug dealers.

My father instilled in me the importance of knowing the intricacies of real life, anticipating that it would benefit me in the future. He shared confidential information, cultivated new friendships easily, and was known for his responsible nature. People sought his help to resolve issues, whether it involved meeting with generals, mayors, or drug dealers. His connections allowed him to navigate the city effortlessly. Even KGB generals preferred his services over their designated drivers. Sometimes, they requested him to read restricted books from foreign countries, which he brought home and discussed with me, fostering my curiosity and interest in the world's complexities.

I had a keen interest in reading about kings, leaders, and those who created empires or left a lasting impact on the world. Particularly intriguing were accounts of the *Ahnenerbe* German space agency during the Nazi era and intelligence agencies. I delved into every book my father brought home, absorbing the knowledge contained within.

My father noticed my enthusiasm and often told me that, once I grew up and turned eighteen, I would go on to study at a KGB university. However, at that time, I was just a boy, yearning for a carefree life filled with joy and fueled by dreams.

As I envisioned my future, I aspired to create something monumental that would leave a mark in the annals of world history. However, inflation was rampant during this period. People began bracing themselves for the impending collapse of the USSR. For the ordinary citizens, it proved to be a devastating blow as the value of money plummeted, wiping out their savings. In an attempt to salvage some semblance of value, the KGB and government leaders endeavored to shift their assets to goods that would retain worth.

The situation escalated as obtaining these goods through regular means was restricted by the omnipresent *Nalogovaya*, or the Federal Taxation Service, Russia's equivalent of the IRS in America, prompting individuals to turn to the black market. The uncertainty about the future created stress, leaving people at a loss about what to expect. Even high-ranking KGB generals found themselves contemplating relocating to other countries or venturing into business. The collapse seemed inevitable. Some approached my father, hoping he could facilitate meetings with mayors or generals to secure funding.

During one such encounter, my father introduced a distraught scientist to a KGB general. As they conversed, I observed the scientist on the verge of tears. However, the general coldly informed him that financial assistance was not forthcoming, emphasizing the need to prioritize his career. It was at this moment that I offered a potential solution, proposing a strategy that could address their predicament.

The strategy involved presenting a scientific project to large companies, showcasing the work and seeking financing, usually 70% of the total cost. Typically, after a thorough review, companies would request additional work and information, delaying financial support. Unbeknownst to the scientist, the company might already be working on a similar scientific project behind the scenes. This is where the KGB general's role became crucial — to manage the company and its partners in other countries. When the company eventually starts selling the technology, the general can monitor its commercial performance, ensuring justice. This approach not only protected the scientist's intellectual property but also enabled the KGB to generate revenue through fines imposed on the company. It was a complex strategy that required careful monitoring and execution, illustrating the intricate dance between scientific innovation, corporate interests, and governmental oversight during a tumultuous time.

My father was elated because, if successful, this strategy would not only aid the scientist but also benefit the general and his other acquaintances. The general decided to proceed with this plan, and for me, it marked my first foray into creation, my inaugural project. I was eager to witness its execution, thrilled that it represented the initial stride toward realizing my dream of contributing something significant to history.

Even though the KGB was aware of my existence, and I, in turn, knew of them, our paths had never crossed directly; they always approached my father. After six months, my father took me with him to an occasion where he and the general celebrated the triumph of the strategy. The general was in high spirits, sharing how the plan had come to fruition and worked seamlessly, resulting in his commendation. He raised numerous toasts in honor of the scientists and enjoyed the revelry.

Throughout the gathering, I couldn't help but wear a constant smile, witnessing the positive outcomes of a strategy that I had played a part in conceptualizing. It was a moment of validation and the realization that even a young individual could contribute to impactful initiatives, setting the stage for my future endeavors.

It was essentially a study of psychology, delving into a deeper understanding of human tendencies towards self-interest, particularly regarding money and career pursuits. The critical insight was that individuals often willingly chose to break the law, acting intentionally, providing fruitful opportunities for various agencies. On this particular night, my father, who had struggled with alcohol-related issues that made him prone to anger, consumed alcohol throughout the night. Contrary to the norm, he was cheerful and engaged in conversation with me. He told me, "Look, my son. Observe, listen, and read – these skills will serve you well in the future." I acknowledged that indeed, these skills would prove beneficial, especially now that I had embarked on a new chapter in America. Despite starting a job in cleaning, my primary focus was on earning money, attending parties, forging new friendships, and establishing connections in my new country.

My cousin generously gifted me one of his cars, and I began working alongside my mother and sister. Every evening, my sister and I attended parties, forging new connections and friendships. Whether with my cousin or newfound friends, each day presented an opportunity to meet different people. We frequently bought alcohol and indulged in the company of others, as I believed that good friendships began with shared experiences over drinks and good food.

My cousin generously gifted me one of his cars, and I began working alongside my mother and sister. Every evening, my sister and I attended parties, forging new connections and friendships. Whether with my cousin or newfound friends, each day presented an opportunity to meet different people. We frequently bought alcohol and indulged in the company of others, as I believed that good friendships began with shared experiences over drinks and good food.

As our circle of friends expanded, my cousin helped me secure a new job in transportation. I became a driver for a Russian company, transporting people to hospitals and back home. This job allowed me to explore the beauty of Minnesota, witnessing its scenic landscapes.

Even though America is a prosperous country where everyone follows the law, my initial impression was of the challenges faced by immigrants. While it's relatively easy to find jobs in smaller companies, such as cleaning roles, the issue arises when employers are reluctant to hire immigrants directly, denying them benefits. Instead, they prefer immigrants to open small companies and work at an hourly rate, often as low as ten dollars, without providing medical or other essential benefits.

Recognizing the hurdles faced by immigrants, I felt compelled to address these challenges if I wanted to create something impactful on a national scale. Life progressed, and my relationships with people improved. I began making numerous friends, entered into a romantic relationship, and even established a company called Zontik LLC, which required dedicated effort to develop further.

Every day brought new social interactions and celebrations, with parties becoming a routine part of life. My older sister arrived in the USA, and we, along with friends, ventured out to clubs for enjoyment, dancing, laughter, and camaraderie. During one such outing, we encountered a family from Duluth, consisting of three brothers and one of their wives. Boris, in particular, stood out with his captivating personality, always ready to lighten the atmosphere with humor. We spent time together, talking, laughing, and enjoying each other's company. It was a delightful experience meeting people from different cities, reinforcing the idea that connections with people worldwide are essential.

This lesson resonated with me, echoing the wisdom of my mother, Lyubov, who had cultivated connections with individuals from around the world. As I navigated my own journey, I appreciated the significance of having a network of people, learning from my mother's ability to build relationships and connections globally.

In the Soviet Union, my mother earned the moniker *"speculianka"*. Business activities were strictly prohibited, yet my mother would travel to Moldavia to procure shoes and gas stoves, which she would then sell. Our visits to New Singereya often involved seeing our grandparents and relatives, and with remarkable ease, my mother forged connections. Her reputation preceded her, especially after the dissolution of the USSR. She played a pivotal role in helping thousands of people establish connections, transforming them from *"perekupchiki"* (speculators) into entrepreneurs. This marked a new era for individuals and the country. The predominant issue following the USSR's breakup was the government's complete elimination of business opportunities, depriving people of the ability to monetize their ideas, creations, or talents.

Even esteemed professionals like scientists, who traditionally received medals and rewards from the political government, were affected. This prompted people to embrace a new government and new ideas, despite the immense challenges of starting life anew. The transition was daunting, but life continued, and everything I learned from my mother and father became an integral part of me — ingrained in my thoughts, experiences, and mindset.

While my mother, sisters, and I embarked on a new life in a different country, my father remained in Kazakhstan, leading his own life with a new family. Yet, the responsibility to care for him remained, recognizing that it's not only the person who raised you throughout your life that deserves care, but also as they age, they become a part of your life that requires attention and support.

Life keeps moving forward, and unexpected turns occur. When my sister arrived, I lost my job, and realizing it wouldn't be a permanent setback, she quickly found me employment in construction, doing siding work. Starting at ten dollars an hour,

I had little choice but to accept. After the first day, the employer observed my efforts and raised my pay to eleven dollars—an improvement, yet a constant reminder of the challenges faced as an immigrant, necessitating employment through a company.

Working hard and putting in overtime became routine, but the reality for many immigrants was the absence of overtime payments and benefits. While it may seem like a positive scenario for the government, fostering the creation of small businesses, individuals often endured a challenging life without essential benefits. Despite the hardships, I found inspiration to contribute to national initiatives through my projects. Utilizing my innovative ideas in various aspects of life, my routine settled into a pattern where each day involved working diligently in the morning, attending parties in the evening, meeting new people, and fostering connections. This rhythm brought new friends into my life, providing a fulfilling and dynamic experience.

The following week, our new friends from Duluth called and invited us to their city. It was a refreshing change to venture beyond the town, unwind, and have a good time. Our friends, who had been residing in the city for an extended period, owned a successful hairdressing salon, having built a career in the new country. As we visited their regular hair shop, we felt a sense of pride, witnessing how our friends had adapted to the system, achieved financial success, and extended an invitation to us.

For me, it was a positive experience—new people, new friendships—and I endeavored to nurture these connections. Life continued to unfold. Over the next six months, our bond deepened, and we made it a weekly routine to visit them. Our visits to Duluth expanded as they introduced us to their friends and the mayor.

However, the reality of an impending economic downturn loomed, and everyone anticipated significant challenges. Despite this, we maintained our weekly gatherings. One day, we broached the topic of the impending depression. I proposed a strategy that, if followed, could help the country generate enough funds to weather the depression, enabling people to make money. Our friends believed in the idea, and we made a promise to present it to the government through the Russian community of Duluth.

When Boris inquired about the cost of the idea, I suggested establishing my corporation (with conditions), retaining all rights, and charging three million for each idea. He found it agreeable, and as we discussed the plan, he asked for the details. I explained that the first step would be a tax release to encourage people to make more money, with the added benefit of receiving money from the government. This concept gained traction and manifested in political campaign discussions.

However, complications arose. Despite my legal refugee status, my idea entered the national political campaign, triggering concerns under Statute 58 of the old Soviet Union law. The CIA swiftly became involved. While it was too late to alter the course, we understood the gravity of the situation. I realized I was under investigation, and the people around me would likely distance themselves. It was a frustrating situation, considering the potential repercussions of merely being associated with me. This, I believed, was one of the most senseless things people could do. The system remained consistent, whether in the USA or Russia, and the unfolding events demonstrated the far-reaching impact of being in the agency's sights.

Life took a dark turn as people I once knew began reaching out, delivering harsh messages: "We don't want to see you, don't call us anymore. You're an outlaw." The turmoil I was experiencing was expected, as I had foreseen these consequences. What mattered most to everyone was merely the name of the idea or project, without understanding the underlying purpose for its creation. The strategy I had devised as a child proved effective, regardless of where I found myself.

During my visit to Duluth the following week, as usual, Boris inquired about the next idea. Despite his recent surgery, I was aware that the CIA was keeping an eye on him. At the border of the country, national security extended its reach to everyone. Boris brought up a topic concerning people selling their organs, and I grasped the potential interest in this concept.

As we discussed the idea, I outlined a plan where individuals who sold their organs would be entitled to ten years of good life (with conditions). The premise was that these individuals would work full-time, frequent bars, restaurants, parks, and consistently enjoy life. Upon their passing, they would donate their organs.

Boris responded positively to the proposal. I anticipated that the project would progress, particularly as organ sellers found it appealing. In general, people are content when provided with a well-paying job, the acquisition of a car and a house, and the chance to find a partner for a joyful life. The prospect of ten years of good life only adds to the appeal.

The organ business is a multitrillion-dollar industry that continues to thrive, especially in countries like Kazakhstan or Russia, where individuals outside major cities often earn a meager fifty dollars a month. In such circumstances, people would willingly embrace the opportunity to extend their lives by another ten years through a program promising a good life.

This program, in essence, serves as a form of medical treatment for individuals grappling with depression or contemplating suicide. The organ sellers become participants in this initiative. However, a significant challenge arises when considering the intervention of skilled psychologists who could inform participants that a decade of a fulfilling life might alleviate their depression and alter their desire to sell organs.

The issue lies in the fact that the stakeholders in the organ business often lack access to psychologists or the ability to monitor the psychological well-being of organ sellers. Consequently, individuals engaged in this business are primarily motivated by the prospect of quick financial gain and tend to disregard ethical or legal considerations. When it comes to obtaining organs, they resort to any means, even if it involves taking them from living individuals or resorting to illegal and potentially lethal methods.

I've already prepared for the next step and am anticipating the moment when they inform me that they won't pay. It's just a matter of time. During our conversation, they asked me a hypothetical question: if someone were to hold a gun to my head, would I choose money or life? It's a common scenario for most people, and the instinctive response is to choose life—a natural reality. However, for me, it was a realization that this situation wasn't unique to me; it would likely happen to everyone. They would employ predatory tactics or any means necessary to avoid payment.

Despite this, it's too late; the game has begun, and it's unstoppable, spreading like a virus. The more people become aware of the potential earnings, the more they will pursue it, especially those who have been preyed upon and left with no alternative. In their circumstances, making money becomes a necessity to ensure they are valued and saved by agencies for whatever reason. I must ensure that my project is genuinely profitable. If everything unfolds as I envision and predict, it will herald a space revolution.

Based on my experience, whenever a project emerges from the government, corporation, or scientists, the first scrutiny comes from potential thieves or opportunists seeking ways to exploit or extract money. To capture their interest in your project, you must be aware of their likely actions and understand the potential pitfalls. Opportunists are always prepared to set traps to achieve their objectives.

Conversely, garnering interest from the government or politicians is a different challenge. To attract their attention, your project must be substantial enough to leave a lasting mark on the country or the world. It should be groundbreaking, capable of making history, and generating significant profits. However, it also entails risks, necessitating careful control of the information disclosed to them. It's crucial to ensure that they are aware of sufficient details for project acceptance and financing for national research and development, without divulging the entirety of the developed and researched project.

From my personal experience, there have been instances where selfish individuals have manipulated situations. In the city of Esik, where I was born, a friend of my father, an executive chief of police, lived his life and performed his duties. One night, I was assaulted by a group of around thirty people, and I suspected his son might have been involved. The next morning, a colleague at work noticed my condition, and I was taken to the police to file a case in a different district. Another chief opened a case promptly.

Later that evening, I received a call from the chief's son, who questioned my actions and mentioned that the chief harbored animosity towards my father. This incident led to numerous calls from friends, indicating that the chief had a strong dislike for the executive chief and was always prepared to set traps for him.

One morning, my father called and shared a cryptic message, mentioning that the chief, who was his friend, would gladly catch his own son to try to eliminate the executive chief of police. He didn't provide any further details. However, when I visited the police department to close the case, I witnessed the chief's growing anger towards me. I explained that my father had instructed me to come. It was then that I truly understood the deep-seated animosity within the department, where colleagues were willing to go to great lengths to eliminate each other for personal gain and promotions.

This realization reinforced the effectiveness of my strategy. Wherever the project was sent, there would always be individuals interested in it, even seeking to replicate it. The game had already begun, and my project had entered the political campaign, unstoppable.

During a visit to Duluth, Boris and I had a crucial conversation. He inquired about payment, expressing uncertainty since everything would belong to me. I replied by highlighting that the decision to dismiss me was made without a full understanding of the project's purpose. He acknowledged his limited influence, stating, "I know, but I'm just a dog. I can recommend you, but it all depends on you." Unspoken thoughts echoed in my mind – not only had they thrown me out, but repercussions could extend to them, especially the person I had shared more with.

Life became challenging swiftly in Golden Valley, where I resided. News spread rapidly, and the police began stopping me for trivial reasons – exceeding the speed limit, riding a bicycle on the road, or discarding cigarette butts on the sidewalk. The community was in an uproar.

Life goes on, and in the political sphere, there's discourse about immigrants, ranging from accusations of taking too much to attempts to distance them from the affluent. However, I sensed a change when a lady, in an interview, expressed gratitude, saying, "Thank you, Georgiy, for bringing food to our family." This statement stirred emotions in the community, leading some people to become angry and focus their attention on me.

Children, on the other hand, were thrilled and exclaimed. However, their parents responded by lightly tapping them on the head, saying things like, "Stupid girl, they don't pay him. If they don't pay him, they won't pay us either." The dynamics of public perception and reactions were becoming apparent.

Everywhere I went, secret signals indicated the affiliation of people—CIA, FBI, or police. Even drug dealers altered their behavior, selling drugs openly in front of me and revealing real prices. My construction job came to a halt as people I knew distanced themselves from me. My sister found me a short-term job with William, helping with some work in his backyard. During our conversation, he asked about my projects, and I introduced him to the concept of a rechargeable car. The next day, he inquired about another idea. I suggested legalizing marijuana and shared prices from *Esik* (Kazakhstan, Moldavia, and Ukraine)—five dollars per kilogram—which could contribute significantly to the budget. He urged me to call Boris.

From that point, everything changed. It became evident that they didn't want to pay me or anyone. For them, acquiring my ideas became important as I wasn't a citizen yet, and according to the law, it's illegal to influence political campaigns. A fierce struggle ensued among people to ensure they would be paid. Everyone became obsessed with knowing the real prices in this country. They all urged me to buy a container, saying, "You've got connections there; we've got connections here." Everyone became interested in making two hundred million from the sale of each container of marijuana. I told them they didn't understand how it works, but I could help if it were for the city. Soon, my ideas began surfacing in government circles.

For them, acquiring my ideas became important as I wasn't a citizen yet, and according to the law, it's illegal to influence political campaigns. A fierce struggle ensued among people to ensure they would be paid. Everyone became obsessed with knowing the real prices in this country. They all urged me to buy a container, saying, "You've got connections there; we've got connections here." Everyone became interested in making two hundred million from the sale of each container of marijuana. I told them they didn't understand how it works, but I could help if it were for the city. Soon, my ideas began surfacing in government circles.

Finally, political candidates acknowledged that clearing this situation would take a long time, and investigations were certain to continue. The agencies were pursuing me aggressively, investigating and recruiting everyone. People went to extremes, breaking the law to ensure they or their children would be investigated. It was chaos, and the agencies lost control. However, I had predicted this; it was bound to happen whether I was involved or not.

I eventually became bankrupt, and without proficiency in English, finding a job was extremely challenging. I had no choice but to go to a shelter. Upon arrival, I was like an angel; everyone knew me there, and all these people remembered me from the times when I had money and would often give them a dollar or something. Despite the difficult times, being hunted, as they say, struggling, I need to do what I planned. I might be going through a hard time right now, but I have to stick to my plan, and I can do it at the library.

The next day, I visited the Minneapolis library, opened an account, and from then on, it became my station. The first thing I needed to do for a space revolution was to create the first draft of the project that would be presented.

Space exploration has been a significant concern for countries over the last fifty years. Each country engages in a race to determine who will be the first to achieve a breakthrough, with leaders vying for the top spot. However, my idea and plan at this time were different. Instead of a competition to determine who would be the first, I proposed a collaborative approach.

The Earth, with all its countries, doesn't have many nations actively involved in space research and financing. It is crucial that they come together with their respective leaders alongside me. It is vital that the government officially opens the project with me and the partners. The significant aspect is the unity that we will achieve together. All rights are mine, and it's only a matter of time before my scientific contributions and projects are recognized officially, designating me as a genius scientist. By law, every genius is deemed to belong to the government. Hence, the "*uslovie*" (condition) was particularly crucial. The real creator has the power to make changes, but not everything will be altered. All that one possesses becomes subject to government oversight through the agreement with me – Georgiy Sergeyevich Garbuz. Similarly, I am bound by the agreement with the government.

I will act as the government's "eye" everything and everybody will be monitored by the eye. Every occurrence will be controlled; it sees, hears, and knows what everybody is doing. Moreover, if anyone violates the law or replicates a similar project or science using my rights, the eye is capable of charging them and taking half of the profit, sharing it with me. It might sound unconventional, but I am the person who believes that everything we have, including ourselves, belongs to the government's eye, and in turn, to the government and ourselves. This transforms every person, encouraging them to work, open businesses with their ideas, get financed, and receive assistance from the government. It makes them a state, federal, country, or international federal employee and a holder of government-partner assets. Wherever they go, they become representatives of the government of the country, and the government must initiate special programs to take care of all these individuals.

Life moved forward, and I began spending more time at the library. The first endeavor I undertook was the development of a business project plan. I initiated the creation of an International Intergalactic Space Corporation, carefully selecting the founders and owners to join me. It became evident that America, Russia, India, Germany, China, England, Kazakhstan, and NASA would serve as the project's founders alongside me, each country receiving a share in the assigned territory around the world.

I delved into the project planning, focusing initially on the Russian territory. Given the chaotic circumstances and the ongoing mess, it was crucial to ensure the safety of children, regardless of whatever might transpire.

I've always had a deep affection for children and a passion for planning. Over the years, I've made thousands of friends, including foster kids. In my youth, I witnessed many of them facing hardships, with some ending up in jail for minor offenses like stealing food or serving in the army. The government seemed indifferent to their struggles, leaving them to fend for themselves. This realization emphasized the importance of including them in my project.

I developed a comprehensive plan to establish Space Schools and Space Academies. The objective was to grant children the opportunity to study in these academies. Recognizing the financial barriers hindering many talented individuals in Russia from pursuing higher education, I aimed to change this through my project, "*Za Detey – For Kids*," within the space initiative. I created the first draft of the project, given the current chaotic situation, and sent it to key authorities, including the special agent in charge of the FBI, the Secretary of Defense, and the government.

This endeavor evolved into a national security concern due to the ongoing mess. To ensure the success of the project, I needed to create something with substantial profit potential. This would guarantee that tax revenue could be directed towards grants for children, ensuring a 100 percent acceptance rate. Despite some health issues I faced, I took it upon myself to research and treat them. My mother had always dreamed of me becoming a doctor, and although that path didn't materialize due to the break-up of the USSR and subsequent disruptions, my knowledge of medical matters became a valuable asset.

I've already submitted my project to government agencies, and now I'm waiting to see how it unfolds, knowing it will turn into a significant game. While residing in a shelter, I dedicated time to researching the lives of black individuals, and the reality I observed was eye-opening. Many engage in drug-related activities, either selling drugs or facilitating street deals and profiting from them. Prostitution is prevalent, with a minimum charge of $20, and they often make around $400 per day. Their lives are intertwined with waiting in line for subsidized apartments and acquiring food stamps, which they often sell.

Despite the challenging circumstances, they support each other, maintaining a community where trust is paramount. Life goes on, and I calculated the potential sales from marijuana in the USA through my project, estimating a minimum of $336 trillion, with half of it generated by black communities. Most individuals involved in the drug trade are unaware of where the money truly goes. Even major drug dealers, when caught and jailed, only represent a fraction of the larger drug money circulation. The impending legalization is creating a frenzy, with people clamoring for supply hoping to make millions.

The general public is caught up in the excitement without understanding the project's purpose or how it should be properly developed. They are driven solely by the desire to make money, overlooking the need for a thorough and correct implementation. The game is underway, as predicted, and the only variable is whether the government will engage in a meaningful conversation with me to discuss the project's details, purpose, and my planned approach. However, given the current trajectory, it appears that chaos, riots, and other disruptions are imminent. The only solution to clear this mess and usher in the Space Revolution is for every country to agree to establish an International Intergalactic Space Federation Government and an International Intergalactic Space Agency worldwide.

I need to ensure that my project plan is ready on time. Transitioning from a shelter to an apartment, my depression has taken a toll on my health. Experiencing severe pain, burning sensations, and sweating in my lungs, I sought medical help. Despite undergoing various tests, the doctors assured me that my lungs were in good condition, but I realized there wasn't a specific medical treatment available for my symptoms. This presented an opportunity to conduct research and incorporate findings into my space project.

Returning home from the doctor, I decided to address the pain in my lungs my way. I opted to administer medication directly to my lungs to ensure effective absorption. Though it involved some risk and had never been attempted before, my depression and pain overshadowed any concerns. I created a medicine, fashioned an inhalation device, and proceeded to the bathroom to inhale the medication through my nose, directing it into my lungs.

Upon inhalation, I immediately felt discomfort in my nose and brain, prompting a runny nose. Enduring the sensation for forty seconds, I eventually decided to blow my nose and did so. Upon the first application, my lungs seemed to undergo a cleansing process, as if a ton of accumulated dirt had been expelled. Intrigued by this initial response, I decided to wait a bit longer to observe further effects. Seated on the couch, I could sense the medicinal molecules permeating through my veins. My veins began to contract, and after five minutes, I felt a significant improvement. It became clear that this experience was crucial for my project, prompting the need for in-depth research.

I headed to the library to draft a comprehensive plan for the scientific aspects of my project. Returning home, I took out my tablet and recorded a video as the genuine creator of the project and its scientific components. I uploaded the video to Russian websites, marking the second phase of my strategy. I strategically placed it on social media platforms, leveraging the freedom of speech without constraints. I waited for reactions. I foresaw a heightened awareness of individuals being targeted under Statute 58, prompting a desire for explanations—someone attempting to undermine them or coerce them into signing contracts under duress.

Executing my strategy successfully required in-depth research on my science works. Week after week, I generated new methods for medical treatment, medications, technology, programs, and projects based on the outcomes of my research. Conducting studies on myself, engaging in social media activities, writing, and creating paintings became my routine, effectively treating my depression. This was crucial, given that seventy percent of the population faces diagnoses of depression, suicidal thoughts, or aggression, which are often inherent in human nature.

Each day brought excitement as I crafted nearly a hundred projects for the space endeavor. Predicting sales and profits for each, I consistently advocated for government investment of tax revenue in grants for foster kids and education in space school academies and space academies. With a plethora of projects and projected sales to cover all aspects, my latest creation was an automatic computer trading program, Cent Boost. This program allowed people to profit from financial markets, earning up to a thousand dollars a day. The program's unique trading methods generated income, benefiting banks, government tax revenue, and individuals. Even for those maintaining regular jobs, they could become space support agents.

Due to my lacking English proficiency, I wrote the project in a combination of Russian and English, a deliberate aspect of my strategy. As the saying goes, if you have a secret and more than three people know about it, it's no longer a secret. Everyone has friends with whom they discuss everything, and there are people who report information. To fully comprehend my project, everyone would need to invite a translator fluent in Russian or English.

While awaiting the progression of my strategy, I conducted private research on all my science works to ensure that they would yield at least 90 percent positive results. My foremost priority was to guarantee effective treatment for all individuals at affordable prices, particularly in countries lacking Medicare where citizens have to bear the costs themselves. My project envisions free healthcare and education for people worldwide if governments accept and legalize marijuana and prostitution globally. It may sound unconventional, but as we, the people, witness these activities occurring illegally, their legalization can bring about benefits for every human being. We, the people, are advocates for a better and prosperous future for the world, and together, we can enact positive change.

As I observed the unfolding events, the USA officially announced the establishment of a Space Command division. Although it is still in its early stages, it deviates from my original plan. The initial plan was to create an international space command because all activities beyond Earth, such as research and colonization of planets, must be collectively undertaken by humanity. Together, we can usher in a new generation, explore Mars and the Moon, and venture into the vast expanse of the universe.

The strategy unfolded as I predicted, creating chaos everywhere, ultimately leading all countries to adapt and establish an International Intergalactical Space Federation government. According to my strategy, the president can only be chosen from former presidents or political leaders of federal governments, selected by people worldwide. The initial candidates included a former president of Kazakhstan, a former president of America, a former chancellor of Germany, and the president of Russia. Many may deem it foolish, thinking I should pursue the presidency for myself, but I prefer to operate in the shadows, orchestrating the game. The leaders of the space government have become international heroes, etched in the annals of international history.

I find joy in making people great, and my commitment to elevating my project is unwavering. Although I am not physically present, my passion for creating and researching ignites when it gains national recognition. I am driven by the desire to contribute something meaningful that aids people around the world.

In the midst of this journey, I lost friends and became a target. Only my mother visits me, and my father communicates with me online. They are not just my parents; they have been my friends, offering care and support in every situation.

I presented my entire project to my father and mother. My mother reacted strongly, expressing concern about potential business-related problems. In contrast, my father, while acknowledging the challenges I might face in securing employment, assured me that everything would be okay. He emphasized that my project could bring hope to millions of people, offering free education and healthcare worldwide through the space project. Financially, he couldn't assist me, but he pledged to leverage his connections. With friends among the ambassadors and officials of Kazakhstan and Russia in the USA, he believed they could support my international project as his son.

He reminded me of the reality I was facing and suggested that the American government should contact these connections. Aware of the difficulties I was encountering, he encouraged me to reveal my full project to the government and advised finding a good job before fully launching the project. His words resonated with me; I needed employment to sustain myself, and I felt a sense of responsibility as my father battled blood cancer, requiring financial assistance that I couldn't provide.

I searched for a job tirelessly, applying online every day wherever opportunities arose. I resided in a neighborhood predominantly inhabited by the black community, which, despite being dangerous at night, had its own unique dynamics. Although I was white, being an immigrant set me apart, and they treated me differently. Interestingly, immigrants formed stronger connections within the black community compared to others. Despite the challenges, there was a sense of camaraderie. Their lives mirrored the struggles faced by the Russian people.

Engaging with the black community, I discovered their authentic way of life. They shared their experiences, including insights into jail life, freely offering advice. I developed a genuine appreciation for the black community, understanding their realities. Despite knowing I was an immigrant, they allowed me to witness and engage in their daily lives. I even learned their language, practicing conversations with them.

In line with my psychological science research, I turned to painting as a form of therapy to alleviate my depression. Interestingly, I managed to sell some of my artworks to one of my doctors.

I had very little money, just enough for alcohol and food, so I began hosting small parties at home. I invited some neighbors, which eventually led to making connections in my neighborhood. They, in turn, introduced me to people from the wider Minneapolis community. Despite the limited funds, it was an opportunity to forge new friendships and establish connections in a new place, and I knew the importance of building relationships.

Finally, I initiated a project to bring a container to the city. While marijuana legalization was on the horizon, I also began preparations for an international intergalactic space-state drug corporation, a government-partnered initiative.

The idea was to encourage drug dealers in every city or state to open legal drug stores in collaboration with the corporation, working on a fifty-fifty partnership. Even with the imminent legalization of marijuana, there was a need for special programs to ensure a smooth transition. I had a project for it: a program where, to become a legal drug dealer, one needed to confess to government agencies as part of an investigation in a special program.

In reality, my idea was for all drug dealers to become federal employees, adhering to government rules and enjoying protection rights, a crucial aspect in a legal government system. This job comes with numerous risks, including assault and robbery, and it was essential for these individuals to be protected by the law. The legalization of marijuana promised not only significant revenue for the government and the people but also the creation of millions of jobs. As soon as it was legalized, both big and small drug dealers could transition to working as employees or businessmen within the legal system, ensuring fair compensation for their work.

As I struggled, I applied for jobs everywhere, reaching out to my friends in the black community. Finally, a hotel hired me as a laundry attendant for $13 per hour, and I was genuinely grateful for the opportunity. However, upon entering the workplace, I noticed that the majority of the staff, including cleaners, were Africans and Mexicans, with only a few Americans in managerial roles. Undeterred, I began my work and research.

In observing my colleagues, I found that Africans and Mexicans were diligent, hardworking individuals who held strong religious beliefs and harbored dreams of achieving the American dream. Over the next few months, I managed to pay for my modest apartment, cover bills, and even host a few gatherings. During a conversation with my black friend, I expressed my newfound understanding of why some people might be hesitant to work. He agreed, emphasizing that being labeled as an "employee" didn't necessarily equate to a comfortable life when the wages barely covered basic necessities.

Despite the hard work, there was a stark reality: a meager income, with only a small amount left for food and other essentials after paying for rent and bills. It became evident that these hardworking individuals could benefit from government assistance, making it more attractive for them to rely on welfare, secure free housing, and receive food assistance, all while engaging in informal cash transactions.

After working for a few more months, my father fell ill and needed medical attention. This prompted my decision to quit my current job and seek better-paying opportunities, understanding the challenges that lay ahead but recognizing the necessity of doing so.

I filled out applications everywhere, seeking opportunities that paid more than twenty-two dollars an hour each day. However, the onset of the COVID-19 pandemic led to widespread job losses, with many people opting to stay home and receive unemployment benefits. Businesses came to a standstill, and even frontline workers decided to leave their jobs, knowing they would be paid to stay home.

This shift in the job market prompted frontline companies to actively seek workers, and I eventually found a job in pizza delivery. The appeal of this job was not just the hourly wage of fifteen dollars but also the additional ten dollars in tips. This allowed me to start saving money diligently, and within the first month, I had saved two thousand dollars.

I felt relieved and shared the news with my father, expressing my joy at finding a job. I reassured him that I could now contribute to covering the costs of his healthcare.

As the country's situation deteriorated, riots erupted everywhere, and chaos ensued. Restrictions on movement were imposed, making it challenging to go outside. This upheaval had been predicted, but its impact was felt across the nation.

I decided to intensify my efforts to save money for the research and development of my project. I found another job in pizza delivery and committed to working seven days a week for ten hours each day. Yes, it was demanding, but the result was significant—I was able to save five thousand dollars every month after covering all my bills.

After a year of hard work, I had saved a total of forty-five thousand dollars. With this substantial amount, I decided to invest in a condo. This decision proved to be a wise move, as it reduced my monthly expenses, making it much easier for me to save money. I continued to strategize on ways to cut costs, such as opting to buy a brand-new car, which lowered my monthly expenses to two hundred eighty-three dollars.

Now, my top priority is saving money for the down payment to secure credit for research and development. I have also begun working on my second project. Thanks to the support of the black community, my English proficiency has significantly improved, enabling me to write my project plan in English and develop a comprehensive strategy to ensure its success at a national level. I dedicate time to this endeavor every day.

I worked diligently, fueled by a plan to bring my project to fruition, meticulously crafted and strategically devised. Currently, I am an American citizen, a first-generation immigrant of American-Russian descent. Despite being half Moldavian, half Ukrainian, and hailing from Kazakhstan, I take pride in being referred to as Russian. It's a part of my identity, and I embrace it fully.

My pursuits extend beyond mere financial gain; they are driven by a commitment to a better future. I am an advocate for the well-being of children worldwide. For me, money serves as a crucial tool—a means to materialize my dreams, to see my ideas resonate globally, and to uplift people's lives in collaboration with the eye and the government.

Amidst the ongoing space revolution, a pivotal chess game, everyone is under the watchful eye of the eagle. I remain steadfast, committed to pushing forward, regardless of the challenges that may unfold in the future.

I strive to secure government financing to ensure that every individual receives fair compensation and enjoys benefits such as free healthcare and education. The unfolding events will reveal how this plan comes to fruition. My project is evolving into an International Intergalactical Space Secret Service Agency, slated to be monitored globally and requiring approval from the Pentagon and the government. According to my design, the head of the space agency is intended to be a former Secretary of Defense. The Pentagon has already received the initial draft of my project.

And they know about me. While awaiting the unfolding of events, I began preparing the Hatorn Project, aiming to legalize prostitution worldwide. This initiative serves various purposes, including boosting the economy of every country and city globally, providing legally accessible relaxation services to the public. The most crucial aspect is to minimize sexual assaults worldwide. This isn't merely about money; according to my scientific research, individuals who have not engaged in sexual activity for more than two months experience sexual depression. Their thoughts become fixated on how to approach and engage in sexual activity, often leading to harassment or assault.

In many cases, especially among the youth, the desire for sex drives them to seek out potential partners. However, they often struggle with initiating conversations about sex, leading to instances of harassment or inappropriate advances. Personally, I find it more effective to engage in open conversations with women about sex, making the interaction more comfortable and consensual. As part of my research, I surveyed a thousand women, asking if they would consider sleeping with a man for a payment of one hundred fifty dollars. Seventy percent responded affirmatively, twenty percent expressed contemplation, and only ten percent declined.

People are ready to seek pleasure, looking for ways to relax and find joy. Often, they resort to illegal activities such as drug use, hiring prostitutes, or excessive drinking as a means of unwinding after work or for various reasons. It's not their fault; the desire for sex is a natural bodily need. Even once a week of engaging in sexual activity can contribute to relaxation and maintaining a healthy mental state, allowing individuals to lead a normal life. They expect it to be easily accessible, and someday, prostitution will likely be legalized. It's only a matter of time before someone recognizes the potential economic benefits and decides to legalize it, understanding that the profits will triple once it becomes a legal and regulated industry.

During my time in downtown Minneapolis, I witnessed the nightly pursuit of sexual encounters for money. It resembled a hunting game, with men approaching women or vice versa. While intriguing to observe, it was also dangerous to live there, with the risk of physical assault or even gunfire. Legalizing prostitution is a step that needs to be taken in the right way. People will find work in the industry, but it is crucial to ensure that they have the right to engage in this profession. This involves proper treatment, fair compensation, and legal protection in case of any issues. For many individuals who have worked in the black market for an extended period, the government requires them to confess before becoming legal state-federal employees through the *Hemken Penance Program.*

The project was meticulously prepared day by day, evolving into a comprehensive plan encompassing almost a hundred different programs, sciences, and projects. It promises substantial and predicted profits. To ensure government approval and enable every country to generate more profits, they will be encouraged to invest in grants through the *Serega Program.* This additional income can be used to provide free healthcare for every individual and cover all military expenses. Given the current global situation, fostering peace between Russia and Ukraine has become a top priority. The *Lyubov* project, centered on love for one's country, people, and children, aims to bring peace and prosperity to every nation. It has the potential to halt inflation, end the suffering and tears of innocent people, and pave the way for a better world where people can thrive once again.

As the project reached its final stage, I submitted it to various government entities, including the Pentagon and other government agencies. Simultaneously, I shared the comprehensive project with hospitals and doctors who had been involved in treating my medical conditions and collaborating on private research related to my scientific endeavors. The goal was to ensure they understood the essence of the project and its potential future impact. Moreover, I aimed to engage these medical professionals as part of my research team. I recognize the importance of having experienced and skilled individuals collaborating on international research, and what you've seen so far is just a glimpse—the project is exponentially larger.

PROJECT LYUBOV

(A Peace Treaty Between Russia and Ukraine in 2024)
Provided by the USA, Germany, and Israel on the Ukrainian
side, and provided by Moldova, Kazakhstan, Belarus, and
the Vatican on the Russian side.

During a war between Russia and Ukraine, both countries have agreed to negotiate a peace treaty to ensure the safety and prosperity of their citizens and the world. As part of the peace treaty, the Russian Federation commits to ensuring the safety of the prisoners of war and provides financial reparations of 100 trillion dollars to Ukraine. The Russian Federation will also engage in the reconstruction of the war-torn cities, including buildings, houses, factories, and other infrastructure in Ukraine.

Furthermore, the Russian Federation will grant citizenship to Ukrainian residents who stay in Crimea. Ukrainian companies' business ownership in Crimea will be restored and come under Russian control. In addition, the Russian Federation will compensate those who were injured during the conflict, including the families of those who lost their lives. All of these efforts are expected to be completed by the year 2030.

According to the terms of the signed peace treaty, Crimea will become part of the Russian Federation. Additionally, as per a peace treaty signed with the mediation of the United States,

Germany, and Israel, the Russian Federation will be entitled to a 5% share in an international intergalactic space corporation.

Moreover, the Russian Federation will also have a 5% ownership stake in four international intergalactic space state drug corporations in Minnesota, Chicago, Washington, and Florida. It will be entitled to a 40% share in a government-partnered international intergalactic space corporation in the Russian Federation and a 10% share in 32 countries designated for Russian territory.

On the Ukrainian side, Ukraine commits to providing healthcare to its citizens, releasing prisoners of war, paying compensation to individuals injured during the war, including the families of those who lost their lives, and seeking UN recognition for its new territory.

This peace treaty, facilitated by Moldova, Kazakhstan, Belarus, and the Vatican, stipulates that Ukraine will be eligible to own a 5% share in each of four international intergalactic space state drug corporations located in Minnesota, Chicago, Washington, and Florida, with Russia holding the majority share at 40%.

Both countries agree to complete the peace treaty and have it signed by the year 2026. The revenue generated from these agreements will be allocated to healthcare, military expenses, space project research programs, education for the world's children, foster care, and other initiatives.

Funding for these educational grants will be derived from taxes on Georgiy S. Garbuz's scientific works and all other projects. The project was initially conceived by Georgiy Sergeyevich Garbuz in 2007 and is being developed under the name "Project For the Future" (G.S. Garbuz) at the National Garbuz Space Academy Trust Corporation.

INTERNATIONAL INTERGALACTICAL SPACE FEDERATION (IISF)

The International Intergalactical Space Federation government is designed to secure and research space. It was founded by the USA, Russia, India, England, Germany, and China, and includes participation from 195 countries around the world, united in their shared vision for Earth's future. As a space government, the IISF is responsible for safeguarding space territory, conducting research, expanding territory through planetary colonization, and seeking existing life across the universe. Their mission also includes improving living conditions on various planets, providing job opportunities and businesses to the planetary populations, among other objectives.

The IISF will collect 6% in federation taxes and 1.89% in sales taxes to support an active space research military program spanning the globe and the galaxy.

The head of the IISF, a President elected for a 10-year term, also serves as the Commander of the Chief International Intergalactic Space Agency. This governance structure includes representation from 195 countries with one Senator from each country, serving 8-year terms. The Space Agency boasts a workforce of 47 million people, supported by 75 billion agents responsible for providing comprehensive information to the Space Government and ensuring the safety and protection of its inhabitants.

The IISF is dedicated to protecting space borders and conducting research. It refrains from participating in local conflicts between countries within planets unless they pose a threat to world or planetary destruction. In such cases, the IISF provides humanitarian aid, offering safety, food transportation, and other forms of support.

The IISF values religious freedom, with Catholic Christianity serving as the main church religion within the federation, offering much-needed spiritual guidance to humanity.

The founder of this visionary project is Georgiy Sergeyevich Garbuz, and the project's revenues will be allocated to healthcare, military agency expenses, space project research, global education initiatives, support for undeserved youth, and the protection of space borders. Funding for educational grants will be sourced from taxes on Georgiy Sergeyevich Garbuz's scientific work and other related projects.

The Project For The Future (G.S. Garbuz) was initiated by Georgiy Sergeyevich Garbuz in 2007 and is being carried forward by the National Garbuz Space Academy Trust Corporation.

INTERNATIONAL INTERGALACTICAL SPACE SECRET SERVICE (IISSS)

Space is the uncharted frontier that humanity has relentlessly sought to explore. Today, millions of people observe the cosmos through telescopes, but regrettably, some individuals exploit this fascination for fraudulent purposes. Every day, millions of people fall victim to these thieves, suffering financial losses and more. Yet, there is no institution equipped to investigate such matters on an international scale and deliver concrete results.

The IISSS was conceived to address this pressing need. Its mission: to safeguard and ensure national stability across planets, space, and countries. The agency's founders hail from five global powers – America, Russia, China, England, India, and Germany. Each country is tasked with the security of the assigned territories in both space and their homelands. With over 17 million active space agents and the support of 7.5 billion citizens, this agency stands poised to provide comprehensive planetary security and deliver critical national safety information.

This agency is dedicated to gathering crucial intelligence and collaborating with other agencies worldwide to prevent the most dangerous threats facing our planet. For instance, in recruiting supporting country space agents, it would require $2.6 trillion annually from the 331.9 million citizens of the USA and $286.8 billion from Russia's 143.4 million supporting agents, with similar calculations for the rest of the world.

Funding for this monumental endeavor can be sourced from dedicated taxes on the legalized marijuana and prostitution industries. The legalization of these practices globally can significantly enhance the safety of our world, our countries, and our people. This approach allows every individual to contribute through their work, provision of information, and compensation. In turn, each country can afford to provide free healthcare, education, and other essential services to its citizens, fostering a cooperative environment for space exploration and planetary colonization.

Key Aspects of the Project:

1. **Financing:** Funding from the legalized marijuana and prostitution industries will support this endeavor.

2. **Selection of Professional Employees and Specialists:** The agency will carefully select and employ experts in relevant fields.

3. **Acquisition of Special Materials and Technologies:** The project will invest in advanced materials and technologies required for its implementation.

4. **Government Oversight:** The establishment of these agencies will be carried out under governmental supervision.

The revenues generated from the legalized industries will be channeled towards the Garbuz Space Academy Program, founded by Georgiy S. Garbuz. This visionary project, initiated in 2007, aims to build the National Garbuz Space Academy and affiliated schools across the globe, providing opportunities for foster care children and underprivileged youth to receive a quality education.

NATIONAL GARBUZ SPACE ACADEMY TRUST CORPORATION

Ownership:

- 40% Georgiy S. Garbuz (Founder)
- 5% Kazakhstan (Government)
- 5% Russian (Government)
- 5% Chinese (Government)
- 5% Germany (European Union Government)
- 5% Indian (Government)
- 5% England (Government)
- 5% United States (Government)
- 5% NASA
- 20% International Intergalactical Space Federation

Territory Responsibilities:

- The Russian Federation is responsible for a space project involving 32 countries, necessitating the presence of security forces, including the military, within this territory.

NATIONAL GARBUZ SPACE ACADEMY TRUST CORPORATION OF IMPERIAL RUSSIA

Ownership:

- 50% National Garbuz Space Academy Trust Corporation (Georgiy Sergeyevich Garbuz)
- 50% National Garbuz Space Academy Trust Corporation of Imperial Russia

PROJECT MAMA

Project MAMA is a comprehensive initiative aimed at the construction and expansion of cities and educational institutions associated with the Garbuz Space Academy. This project focuses on the care, education, training, and future prospects of orphaned children (AA, A1, A2) and children from military backgrounds (A3) in Russia and the CIS countries.

These children will have opportunities for living, studying, and participating in a range of projects, including "Children of the Future," "MAMA," "Brother," "Vatican," and "Illiaja." Furthermore, the project aims to ensure the safety and protection of all participants.

In later phases, some participants may have the opportunity for relocation to well-to-do European countries such as England, Germany, Spain, Greece, with exceptions for certain prohibited countries in Africa and the Middle East, including Saudi Arabia.

Another phase involves potential relocation to Antarctica for work, education, training, and participation in projects like "Antar 5," "Children of the Future," "MAMA," "Brother," "Vatican," and others.

Additionally, there is a provision for potential relocation to European countries for continued work, education, and training of project participants. Special considerations and monitoring are in place for the protection of these children, particularly in the event of unforeseen circumstances.

NATIONAL GARBUZ SPACE ACADEMY TRUST CORPORATION OF KAZAKHSTAN

Ownership:

- 50% National Garbuz Space Academy Trust Corporation (Georgiy Sergeyevich Garbuz)
- 40% National Garbuz Space Academy Trust Corporation of Kazakhstan
- 10% National Garbuz Space Academy Trust Corporation of Imperial Russia

PROJECT PAPA

Expansion and development of the city of Esik, along with the establishment of Garbuz Space School Academies and Garbuz Space Academies, involving the relocation and integration of orphaned children (AA) (A1) (A2) and children of military personnel in Kazakhstan (A3) (inclusive – to be personally verified).

The purpose is to provide these individuals with living arrangements, education, upbringing, training, work opportunities, and involvement in projects such as "PAPA," "Children of the Future," and "MAMA."

The subsequent plan involves their transfer to affluent countries, like England, Germany, Spain, and other European countries. The relocation is intended for further education, training, work, and participation in projects like "Children of the Future," "Brother," "Vatican," and "Illiaja," ensuring the protection of the participants.

Additionally, a phase involves their transfer to Antarctica for work, education, training, and participation in projects such as "Antar 5," "Children of the Future," "MAMA," "Brother," "Vatican," and "PAPA."

Special supervision is enforced for the children, and in the event of any harm, the legal provision "For Children" under Article 58, including charges of terrorism, confiscation of both registered and unregistered property, and execution with a special cartridge, is applicable.

NATIONAL GARBUZ SPACE ACADEMY TRUST CORPORATION OF CHECHNYA

Ownership:

- 50% National Garbuz Space Academy Trust Corporation (Georgiy Sergeyevich Garbuz)
- 40% National Garbuz Space Academy Trust Corporation of Chechnya
- 10% National Garbuz Space Academy Trust Corporation of Imperial Russia

PROJECT CHECHENECH

Enrollment of orphaned children (A1) from Chechnya, Russia, both before and after their integration into Russia, is undertaken for the purposes of education, upbringing, training, work, and participation in Project MAMA. Subsequently, there is a planned transfer to specific countries, excluding those forbidden (Africa, East, except Saudi Arabia and Iran), with the aim of participating in Project MAMA. This initiative also ensures the protection of participants in projects such as "Children of the Future" and "MAMA," with strict measures in place, including the use of a special cartridge in case of any threats.

NATIONAL GARBUZ SPACE ACADEMY TRUST CORPORATION OF BELARUS

Ownership:

- 50% National Garbuz Space Academy Trust Corporation (Georgiy Sergeyevich Garbuz)
- 40% National Garbuz Space Academy Trust Corporation of Belarus
- 10% National Garbuz Space Academy Trust Corporation of Imperial Russia

PROJECT POTATO

The development and expansion of a city in Belarus (Russia) involve the establishment of Garbuz Space School Academies and Garbuz Space Academies, with a focus on utilizing orphaned children (A1) from Belarus for their education, upbringing, training, work, and participation in various projects, including "MAMA," "Potato," "Chernovtsi," and others associated with the overarching initiative "Children of the Future."

Following this phase, there is a planned transfer of these individuals to affluent European countries (specific countries to be determined), excluding prohibited regions such as Africa and the East. The primary objective is their continued participation in the "MAMA" project, along with ensuring the safety and protection of participants engaged in both the "MAMA" and "Children of the Future" projects.

NATIONAL GARBUZ SPACE ACADEMY TRUST CORPORATION OF UZBEKISTAN

Ownership:

- 50% National Garbuz Space Academy Trust Corporation (Georgiy Sergeyevich Garbuz)
- 40% National Garbuz Space Academy Trust Corporation of Uzbekistan
- 10% National Garbuz Space Academy Trust Corporation of Imperial Russia

PROJECT AMU DARYA

The undertaking involves the construction and expansion of a city in Uzbekistan (Russia), employing orphaned children (A1) from Uzbekistan, both before and after joining Russia. The objective is to provide them with education, upbringing, training, work opportunities, and engagement in various projects such as "Amu Darya," "PAPA," "MAMA," "Potato," "Chernivtsi," and other initiatives associated with the overarching theme of "Children of the Future."

Following this phase, there is a planned transfer of these individuals to well-established countries like Turkey and Egypt, with specific countries to be determined. Notably, prohibited regions such as Africa are excluded. The purpose of the transfer is to facilitate their participation in projects like "Pharaoh," "Children of the Future," "PAPA," the "Mulla" Project, while ensuring the safety and protection of participants engaged in projects like "MAMA," "Children of the Future," and "PAPA."

The children are placed under special supervision, anticipating any unforeseen events. Legal provisions, including Article 58 with charges of terrorism, confiscation of both registered and unregistered property, and execution with a special cartridge, are applicable in case of harm to the children.

NATIONAL GARBUZ SPACE ACADEMY TRUST CORPORATION OF KYRGYZSTAN

Ownership:

- 50% National Garbuz Space Academy Trust Corporation (Georgiy Sergeyevich Garbuz)
- 40% National Garbuz Space Academy Trust Corporation of Kyrgyzstan
- 10% National Garbuz Space Academy Trust Corporation of Imperial Russia

PROJECT ISSYK-KUL

The initiative involves the construction and expansion of a city in Kyrgyzstan (Russia), establishing Garbuz Space School Academies and Garbuz Space Academies, utilizing orphaned children (A1) from Kyrgyzstan, both before and after joining Russia. The primary aim is to provide these children with education, upbringing, training, work opportunities, and involvement in projects such as "MAMA," "Issyk-Kul," "Amu Darya," "Illayazha," "Potato," "Chernivtsi," and other initiatives associated with the overarching theme of "Children of the Future."

Subsequently, there is a planned transfer of these individuals to well-established countries like India, Turkey, and Egypt, with specific countries to be determined. Notably, excluded regions include prohibited areas such as Africa. The purpose of the transfer is to facilitate their participation in projects like "Pharaoh," "Children of the Future," "MAMA," the "Mulla" Project, while ensuring the safety and protection of participants engaged in projects like "Mom," "Children of the Future," and "PAPA."

The children are placed under special supervision to address any potential risks. Legal provisions, including Article 58 with charges of terrorism, confiscation of both registered and unregistered property, and execution with a special cartridge, are applicable in the event of harm to the children.

NATIONAL GARBUZ SPACE ACADEMY TRUST CORPORATION OF AZERBAIJAN

Ownership:

- 50% National Garbuz Space Academy Trust Corporation (Georgiy Sergeyevich Garbuz)
- 40% National Garbuz Space Academy Trust Corporation of Azerbaijan
- 10% National Garbuz Space Academy Trust Corporation of Imperial Russia

PROJECT GYZ GALASY

The endeavor encompasses the construction and expansion of a city in Azerbaijan (Russia), involving the utilization of orphaned children (A1) from Azerbaijan, both before and after their integration into Russia. The primary objective is to provide these children with education, upbringing, training, work opportunities, and engagement in projects such as "Gyz Galasy," "MAMA," "Issyk-Kul," "Amu Darya," "PAPA," "Potato," "Chernovtsi," and other initiatives related to the overarching theme of "Children of the Future."

Following this phase, there is a planned transfer of these individuals to countries like India, Iran, Egypt, and Indonesia, with specific countries to be determined. Notably, excluded regions include forbidden areas such as Africa. The purpose of the transfer is to facilitate their participation in projects like "Pharaoh," "Children of the Future," "MAMA," "Mulla," "Gyz Galasy," and to ensure the protection of participants engaged in projects like "MAMA," "Children of the Future," "Illayazha."

Subsequently, there is another planned transfer to the countries of Brazil, where they will participate in the project and work, ensuring the continued protection of participants engaged in projects like "Mom," "Children of the Future," "PAPA," and "Hunny." The children remain under special supervision in case of unforeseen events, with legal provisions such as Article 58 in place, including charges of terrorism, confiscation of both registered and unregistered property, and execution with a special cartridge, in the event of harm to the children.

NATIONAL GARBUZ SPACE ACADEMY TRUST CORPORATION OF MONGOLIA

Ownership:

- 50% National Garbuz Space Academy Trust Corporation (Georgiy Sergeyevich Garbuz)
- 40% National Garbuz Space Academy Trust Corporation of Mongolia
- 10% National Garbuz Space Academy Trust Corporation of Imperial Russia
-

PROJECT GOBI

The development and expansion of a city in Mongolia (Russia) involve the utilization of orphaned children (A1) from Mongolia, both before and after their integration into Russia. The primary purpose is to provide these children with education, upbringing, training, work opportunities, and involvement in projects such as the Gobi project, "Giz Galasy," "MAMA," "Issyk-Kul," "Amu Darya," "PAPA," "Potato," "Chernovtsi," and other initiatives related to the overarching theme of the "Children of the Future" project.

Following this phase, there is a planned transfer of these individuals to countries like India, Iran, Egypt, and Indonesia, with specific countries to be determined. Notably, excluded regions include forbidden areas such as Africa. The purpose of the transfer is to facilitate their participation in projects like "Pharaoh," "MAMA," the "Mulla," "Gyz Galasy" project, and to ensure the protection of participants engaged in projects like "MAMA," "Children of the Future," "PAPA."

Subsequently, there is another planned transfer to the countries of India and China, where they will participate in the project and work, ensuring the continued protection of participants engaged in projects like "MAMA," "Children of the Future," "PAPA," "Hunny," and "Jochi." The children remain under special supervision in case of unforeseen events, with legal provisions such as Article 58 in place, including charges of terrorism, confiscation of both registered and unregistered property, and execution with a special cartridge, in the event of harm to the children.

NATIONAL GARBUZ SPACE ACADEMY TRUST CORPORATION OF TURKMENISTAN

Ownership:

- 50% National Garbuz Space Academy Trust Corporation (Georgiy Sergeyevich Garbuz)
- 40% National Garbuz Space Academy Trust Corporation of Turkmenistan
- 10% National Garbuz Space Academy Trust Corporation of Imperial Russia

PROJECT DEHISTAN

The undertaking involves the construction and expansion of a city in Turkmenistan (Russia), incorporating the use of orphaned children (A1) from Turkmenistan, both before and after their integration into Russia. The primary focus is to provide these children with training, education, upbringing, work opportunities, and participation in projects such as "Dehistan," "Gobi," "Gyz Galasy," "MAMA," "Issyk-Kul," "Amu Darya," "PAPA," "Potato," "Chernovtsi," and other initiatives linked to the overarching theme of the "Children of the Future" project.

Following this phase, there is a planned transfer of these individuals to countries such as Turkey, Iran, Egypt, and Indonesia. The purpose of the transfer is to facilitate their involvement in projects like "Pharaoh," "MAMA," the "Mulla," and the "Gyz Galasy" project, ensuring the protection of participants engaged in projects like "MAMA," "Children of the Future," and "Illayazha."

Subsequently, there is another planned transfer to the countries of India and China, where they will participate in the project and work, while ensuring the continued protection of participants engaged in projects like "MAMA," "Children of the Future," "PAPA," "Hunny," and "Jochi." The children remain under special supervision in case of unforeseen events, with legal provisions such as Article 58 in place, including charges of terrorism, confiscation of both registered and unregistered property, and execution with a special cartridge, in the event of harm to the children.

NATIONAL GARBUZ SPACE ACADEMY TRUST CORPORATION OF ARMENIA

Ownership:

- 50% National Garbuz Space Academy Trust Corporation (Georgiy Sergeyevich Garbuz)
- 40% National Garbuz Space Academy Trust Corporation of Armenia
- 10% National Garbuz Space Academy Trust Corporation of Imperial Russia

PROJECT CASCADE

The undertaking involves the construction and expansion of a city in Armenia (Russia), establishing Garbuz Space School Academies and Garbuz Space Academies, with the incorporation of orphaned children (A1) from Armenia, both before and after their integration into Russia. The primary goal is to provide these children with education, upbringing, training, work opportunities, and participation in projects such as "MAMA," "PAPA," and other initiatives associated with the overarching theme of "Children of the Future."

Following this phase, there is a planned transfer of these individuals to countries like Portugal and other European nations. Notably, excluded regions include forbidden areas such as Africa. The purpose of the transfer is to facilitate their participation in projects like "Mom," "Pharaoh," ensuring the protection of participants engaged in projects like "Mom," "Children of the Future," and "Illayazh."

Subsequently, there is another planned transfer to the countries of India and China, where they will participate in the project and work, while ensuring the continued protection of participants engaged in projects like "MAMA," "Children of the Future," "PAPA," "Hunny," and "Jochi." The children remain under special supervision in case of unforeseen events, with legal provisions such as Article 58 in place, including charges of terrorism, confiscation of both registered and unregistered property, and execution with a special cartridge, in the event of harm to the children.

NATIONAL GARBUZ SPACE ACADEMY TRUST CORPORATION OF GEORGIA

Ownership:

- 50% National Garbuz Space Academy Trust Corporation (Georgiy Sergeyevich Garbuz)
- 40% National Garbuz Space Academy Trust Corporation of Georgia
- 10% National Garbuz Space Academy Trust Corporation of Imperial Russia

PROJECT KHERTVISI

The construction and expansion of a city in Georgia (Russia), featuring Garbuz Space School Academies and Garbuz Space Academies, involve the integration of orphaned children (A1) from Georgia, both before and after joining Russia. The primary purpose is to provide these children with education, upbringing, training, and future work opportunities, participating in projects such as "Khertvisi," "MAMA," "PAPA," and other initiatives related to the overarching theme of "Children of the Future."

Following this phase, there is a planned transfer of these individuals to countries like Norway and other European nations. Notably, excluded regions include forbidden areas such as Africa. The purpose of the transfer is to facilitate their participation in the project, particularly "MAMA," ensuring the protection of participants engaged in projects like "MAMA," "Children of the Future," and "Illayazh."

Subsequently, there is another planned transfer to the countries of India and China, where they will participate in the project and work, while ensuring the continued protection of participants engaged in projects like "MAMA," "Children of the Future," "PAPA," "Hunny," and "Jochi." The children remain under special supervision in case of unforeseen events, with legal provisions such as Article 58 in place, including charges of terrorism, confiscation of both registered and unregistered property, and execution with a special cartridge.

Additionally, it should be noted that 32 countries, including Georgia, are assigned to Russia in a specified territory. In the context of a space project, the United States is assigned 32 countries, encompassing a total of 197 countries. In this expansive space project, space security is deemed crucial, and the designated territory for the United States involves collaboration with 32 countries, emphasizing the need for robust space security measures.

NATIONAL GARBUZ SPACE ACADEMY TRUST CORPORATION OF UNITED STATES OF AMERICA

Ownership:

- 50% National Garbuz Space Academy Trust Corporation (Georgiy Sergeyevich Garbuz)
- 50% National Garbuz Space Academy Trust Corporation of the United States of America

PROJECT HUNNY

Construction and expansion of cities in America, specifically Rockford, MN, involve the establishment of Garbuz Space School Academies and Garbuz Space Academies, incorporating the use and relocation of orphaned children (A1) from Kazakhstan, Russia, and various united countries. The objective is to provide these children with education, upbringing, training, work opportunities, and engagement in projects such as "Children of the Future," "MAMA," "PAPA," and other initiatives related to the Beauty project.

Following this phase, there is a planned transfer of these individuals to Antarctic countries, including England, Germany, and other European nations. Notably, excluded regions encompass forbidden areas such as Africa and the East. The purpose of the transfer is to facilitate their participation in the "Children of the Future" project and ensure the protection of participants in the America project.

Children remain under special supervision in case of unforeseen events, with legal provisions such as Article 58 in place, including charges of terrorism, confiscation of both registered and unregistered property, and execution with a special cartridge if harm befalls them.

NATIONAL GARBUZ SPACE ACADEMY TRUST CORPORATION OF CANADA

Ownership:

- 50% National Garbuz Space Academy Trust Corporation (Georgiy Sergeyevich Garbuz)
- 40% National Garbuz Space Academy Trust Corporation of Canada
- 10% National Garbuz Space Academy Trust Corporation of the United States of America

PROJECT TORONTO

Construction and development of cities in Canada involve the establishment of Garbuz Space School Academies and Garbuz Space Academies, incorporating the use and relocation of orphaned children (A1) from Kazakhstan, Russia, and various united countries. The primary aim is to provide these children with education, upbringing, training, work opportunities, and involvement in projects such as "Children of the Future," "MAMA," "PAPA," and other initiatives related to the Beauty project.

Subsequently, there is a planned transfer of these individuals to Antarctic countries, including England, Germany, and other European nations. Notably, excluded regions encompass forbidden areas such as Africa and the East. The purpose of the transfer is to facilitate their participation in the "Children of the Future" project and ensure the protection of participants in the Canada project.

Children remain under special supervision in case of unforeseen events, with legal provisions such as Article 58 in place. This includes charges of terrorism, confiscation of both registered and unregistered property, and execution with a special cartridge if harm befalls them.

NATIONAL GARBUZ SPACE ACADEMY TRUST CORPORATION OF MEXICO

Ownership:

- 50% National Garbuz Space Academy Trust Corporation (Georgiy Sergeyevich Garbuz)
- 40% National Garbuz Space Academy Trust Corporation of Mexico
- 10% National Garbuz Space Academy Trust Corporation of the United States of America

PROJECT AZTEC

Construction and expansion of cities in Mexico involve the establishment of Garbuz Space School Academies and Garbuz Space Academies, incorporating the use and relocation of orphaned children (A1) from Mexico, El Salvador, and various united countries. The primary goal is to provide these children with education, upbringing, training, work opportunities, and involvement in projects such as "Children of the Future," "MAMA," "PAPA," and other initiatives related to the Hurtigruten project.

Subsequently, there is a planned transfer of these individuals to Antarctic countries, including England, Germany, and other European nations. Notably, excluded regions encompass forbidden areas such as Africa and the East. The purpose of the transfer is to facilitate their participation in the "Children of the Future" project and ensure the protection of participants in the Mexico project.

Children remain under special supervision in case of unforeseen events, with legal provisions such as Article 58 in place. This includes charges of terrorism, confiscation of both registered and unregistered property, and execution with a special cartridge if harm befalls them.

NATIONAL GARBUZ SPACE ACADEMY TRUST CORPORATION OF ARGENTINA

Ownership:

- 50% National Garbuz Space Academy Trust Corporation (Georgiy Sergeyevich Garbuz)
- 40% National Garbuz Space Academy Trust Corporation of Argentina
- 10% National Garbuz Space Academy Trust Corporation of the United States of America

PROJECT ROSARIO

Space School Academies and Garbuz Space Academies, incorporating the use and relocation of orphaned children (A1) from Mexico, El Salvador, and various united countries. The primary objective is to provide these children with education, upbringing, training, work opportunities, and involvement in projects such as "Children of the Future," "MAMA," "PAPA," and other initiatives related to the Hurtigruten project.

Subsequently, there is a planned transfer of these individuals to Antarctic countries, including England, Germany, and other European nations. Notably, excluded regions encompass forbidden areas such as Africa and the East. The purpose of the transfer is to facilitate their participation in the "Children of the Future" project and ensure the protection of participants in the Argentina project.

Children remain under special supervision in case of unforeseen events, with legal provisions such as Article 58 in place. This includes charges of terrorism, confiscation of both registered and unregistered property, and execution with a special cartridge if harm befalls them.

NATIONAL GARBUZ SPACE ACADEMY TRUST CORPORATION OF BRAZIL

Ownership:

- 50% National Garbuz Space Academy Trust Corporation (Georgiy Sergeyevich Garbuz)
- 40% National Garbuz Space Academy Trust Corporation of Brazil
- 10% National Garbuz Space Academy Trust Corporation of the United States of America

PROJECT AMAZONAS

Construction and expansion of cities in Brazil involve the establishment of Garbuz Space School Academies and Garbuz Space Academies, incorporating the use and relocation of orphaned children (A1) from Mexico, El Salvador, and various united countries. The primary objective is to provide these children with education, upbringing, training, work opportunities, and involvement in projects such as "Children of the Future," "MAMA," "PAPA," and other initiatives related to the Amazonas project.

Subsequently, there is a planned transfer of these individuals to Antarctic countries, including England, Germany, and other European nations. Notably, excluded regions encompass forbidden areas such as Africa and the East. The purpose of the transfer is to facilitate their participation in the "Children of the Future" project and ensure the protection of participants in the Brazil project.

Children remain under special supervision in case of unforeseen events, with legal provisions such as Article 58 in place. This includes charges of terrorism, confiscation of both registered and unregistered property, and execution with a special cartridge if harm befalls them.

NATIONAL GARBUZ SPACE ACADEMY TRUST CORPORATION OF PERU

Ownership:

- 50% National Garbuz Space Academy Trust Corporation (Georgiy Sergeyevich Garbuz)
- 40% National Garbuz Space Academy Trust Corporation of Peru
- 10% National Garbuz Space Academy Trust Corporation of the United States of America

PROJECT CARAL

Construction and expansion of cities in Peru involve the establishment of Garbuz Space School Academies and Garbuz Space Academies, incorporating the use and relocation of orphaned children (A1) from Mexico, El Salvador, and various united countries. The primary objective is to provide these children with education, upbringing, training, work opportunities, and involvement in projects such as "Children of the Future," "MAMA," "PAPA," and other initiatives related to the Caral project.

Subsequently, there is a planned transfer of these individuals to Antarctic countries, including England, Germany, and other European nations. Notably, excluded regions encompass forbidden areas such as Africa and the East. The purpose of the transfer is to facilitate their participation in the "Children of the Future" project and ensure the protection of participants in the Caral project.

Children remain under special supervision in case of unforeseen events, with legal provisions such as Article 58 in place. This includes charges of terrorism, confiscation of both registered and unregistered property, and execution with a special cartridge if harm befalls them.

NATIONAL GARBUZ SPACE ACADEMY TRUST CORPORATION OF CUBA

Ownership:

- 50% National Garbuz Space Academy Trust Corporation (Georgiy Sergeyevich Garbuz)
- 40% National Garbuz Space Academy Trust Corporation of Cuba
- 10% National Garbuz Space Academy Trust Corporation of the United States of America

PROJECT HAVANA

Construction and expansion of cities in Peru involve the establishment of Garbuz Space School Academies and Garbuz Space Academies, incorporating the use and relocation of orphaned children (A1) from Cuba, El Salvador, and various united countries. The primary objective is to provide these children with education, upbringing, training, work opportunities, and involvement in projects such as "Children of the Future," "MAMA," "PAPA," and other initiatives related to the Havana project.

Subsequently, there is a planned transfer of these individuals to Antarctic countries, including England, Germany, and other European nations. Notably, excluded regions encompass forbidden areas such as Africa and the East. The purpose of the transfer is to facilitate their participation in the "Children of the Future" project and ensure the protection of participants in the Havana project.

Children remain under special supervision in case of unforeseen events, with legal provisions such as Article 58 in place. This includes charges of terrorism, confiscation of both registered and unregistered property, and execution with a special cartridge if harm befalls them.

NATIONAL GARBUZ SPACE ACADEMY TRUST CORPORATION OF DOMINICAN REPUBLIC

Ownership:

- 50% National Garbuz Space Academy Trust Corporation (Georgiy Sergeyevich Garbuz)
- 40% National Garbuz Space Academy Trust Corporation of Dominican Republic
- 10% National Garbuz Space Academy Trust Corporation of the United States of America

PROJECT SANTO DOMINGO

Construction and expansion of cities in Peru involve the establishment of Garbuz Space School Academies and Garbuz Space Academies, incorporating the use and relocation of orphaned children (A1) from the Dominican Republic, El Salvador, and various united countries. The primary objective is to provide these children with education, upbringing, training, work opportunities, and involvement in projects such as "Children of the Future," "MAMA," "PAPA," and other initiatives related to the Santo Domingo project.

Subsequently, there is a planned transfer of these individuals to Antarctic countries, including England, Germany, and other European nations. Notably, excluded regions encompass forbidden areas such as Africa and the East. The purpose of the transfer is to facilitate their participation in the "Children of the Future" project and ensure the protection of participants in the Santo Domingo project.

Children remain under special supervision in case of unforeseen events, with legal provisions such as Article 58 in place. This includes charges of terrorism, confiscation of both registered and unregistered property, and execution with a special cartridge if harm befalls them.

NATIONAL GARBUZ SPACE ACADEMY TRUST CORPORATION OF VENEZUELA

Ownership:

- 50% National Garbuz Space Academy Trust Corporation (Georgiy Sergeyevich Garbuz)
- 40% National Garbuz Space Academy Trust Corporation of Venezuela
- 10% National Garbuz Space Academy Trust Corporation of the United States of America

PROJECT CARACAS

Construction and expansion of cities in Peru involve the establishment of Garbuz Space School Academies and Garbuz Space Academies, incorporating the use and relocation of orphaned children (A1) from Venezuela, El Salvador, and various united countries. The primary objective is to provide these children with education, upbringing, training, work opportunities, and involvement in projects such as "Children of the Future," "MAMA," "PAPA," and other initiatives related to the Caracas project.

Subsequently, there is a planned transfer of these individuals to Antarctic countries, including England, Germany, and other European nations. Notably, excluded regions encompass forbidden areas such as Africa and the East. The purpose of the transfer is to facilitate their participation in the "Children of the Future" project and ensure the protection of participants in the Caracas project.

Children remain under special supervision in case of unforeseen events, with legal provisions such as Article 58 in place. This includes charges of terrorism, confiscation of both registered and unregistered property, and execution with a special cartridge if harm befalls them.

Additionally, it's noteworthy that 32 countries, including Venezuela and others, are assigned to the United States territory. In this context, the assigned territory for the United States encompasses a total of 32 countries, including 197 countries in a collaborative effort.

NATIONAL GARBUZ SPACE ACADEMY TRUST CORPORATION OF INDIA

Ownership:

- 50% National Garbuz Space Academy Trust Corporation (Georgiy Sergeyevich Garbuz)
- 40% National Garbuz Space Academy Trust Corporation of India
- 10% National Garbuz Space Academy Trust Corporation of the United States of America

PROJECT COLKATA

The initiative involves constructing and expanding cities and Garbuz Space Academies in India, specifically designed for the inclusion and integration of orphaned children (AA, A1, and A2), children of military personnel (A3), and those from Russia and the CIS countries (verification inclusive). The purpose is to provide a living environment, education, training, and participation in various projects such as "Children of the Future," "Mom," "Illiaja," and others associated with the "Children of the Future" initiative.

Furthermore, the plan includes the subsequent transfer of participants to affluent countries like England, Germany, Spain, Greece, and other European nations. Note that countries in Africa and the East, excluding Saudi Arabia, are prohibited.

Participants will actively engage in projects such as "Children of the Future," "MAMA," "Brother," "Vatican," "Illiaja," with the objective of ensuring the protection and well-being of all involved. Subsequently, there are plans for their transfer to Antarctica, where they will engage in work, education, training, and participation in projects like "Antar 5."

Additionally, participants will be relocated to European countries for continued work, education, and training. It is crucial to highlight that children under special supervision will be monitored closely, and in case of any untoward incidents, severe legal consequences, such as the application of Statute 58 on terrorism, confiscation of registered and unregistered property, and execution with a special cartridge, will be enforced.

PROJECT HERAT

The initiative involves the expansion and construction of cities in India, along with the establishment of Garbuz Space School Academies and Garbuz Space Academies. This includes the utilization and relocation of orphaned children (AA, A1, and A2) and children of the military in India (A3).

The primary objectives are to provide living arrangements, education, upbringing, training, work, and participation in projects such as "PAPA," "Children of the Future," and "MAMA." Following this phase, there is a planned transfer to prosperous countries like England, Germany, Spain, and other European nations (with exceptions explicitly noted).

Once in these countries, participants will engage in further education, training, work, and participation in projects such as "Children of the Future," "Brother," "Vatican," and "Illiaja," with a focus on ensuring the protection of all involved. Subsequently, there is a planned transfer to Antarctica for continued work, education, training, and participation in projects like "Antar 5," "Children of the Future," "MAMA," "Brother," "Vatican," and "PAPA" (among others).

It is crucial to highlight that children under special supervision will be monitored closely, and in case of any untoward incidents, severe legal consequences, such as the application of Statute 58 on terrorism, confiscation of registered and unregistered property, and execution with a special cartridge, will be enforced.

PROJECT KARAJ

The plan involves the construction and expansion of cities and Garbuz Space Academies in Iran, focusing on the utilization and relocation of orphaned children (AA) (A1) (A2) and children of the military (A3). This will be thoroughly verified for Russia and the CIS countries.

The objectives include providing a place for living, passing, studying, educating, training, working, and participating in projects such as "Children of the Future," "MAMA," "Illiaja," and others associated with the "Children of the Future" initiative. Subsequently, there's a planned transfer of participants to affluent countries like England, Germany, Spain, and Greece, among other European nations. (Note: Prohibited countries include those in Africa and the East, except Saudi Arabia.)

Participants will engage in projects like "Children of the Future," "MAMA," "Brother," "Vatican," and "Illiaja," ensuring the protection of project participants. Another phase includes the transfer to Antarctica for work, education, training, and participation in projects like "Antar 5," "Children of the Future," "MAMA," "Brother," "Vatican," and "PAPA" (among others).

Further transfers to European countries are planned for work, education, training, with participants under special supervision in case of unforeseen incidents. Legal consequences, including the application of Article 58 on terrorism, confiscation of registered and unregistered property, and execution with a special cartridge, are stipulated under the sentence "For Children."

PROJECT PANIJAB

The initiative involves the expansion and development of the city of Esik, including the establishment of Garbuz Space Schools and Garbuz Space Academies. The plan includes the utilization and relocation of orphaned children (AA) (A1) (A2) and children of the military of Kazakhstan (A3), with a commitment to personally verify all aspects.

The primary goals encompass providing a living environment, education, upbringing, training, work, and participation in projects such as "PAPA," "Children of the Future," and "MAMA." There are future plans for transferring participants to prosperous countries like England, Germany, Spain, and other European nations, with certain countries being explicitly forbidden.

For the educational, training, and work-related involvement in projects such as "Children of the Future," "Brother," "Vatican," and "Illiaja," as well as ensuring the protection of project participants, further transfers are considered. Additionally, there's a subsequent transfer to Antarctica for work, education, training, and participation in projects like "Antar 5," "Children of the Future," "MAMA," "Brother," "Vatican," and "PAPA" (among others).

Children participating in these projects will be under special supervision in case of unforeseen incidents, with the application of legal consequences outlined in Article 58 on terrorism. This includes the potential confiscation of registered and unregistered property and execution using a special cartridge.

PROJECT AHTAP

The proposal involves the creation and design of the Space Business Center or "AHTAP," dedicated to achieving a permanent flight around Earth, hosting tourists, overseeing the sale of top-secret technologies, providing technological support for spacecraft exploration in outer space, securing Earth's boundaries, and preventing potential comet impacts.

The vastness of space remains a mysterious frontier for humanity that warrants exploration. Currently, human space exploration relies on satellites or individual spacecraft to the moon, searching for natural minerals or signs of life. However, the secrets of space persist. Georgiy S. Garbuz's project "Lyubov," aimed at establishing an International Intergalactical Space Government, has been presented for government review. If approved within the next ten years, the project envisions deploying 5000 large space liners for research on Mars, the Moon, or other planets. This ambitious undertaking necessitates the creation of a space base, leading to the conceptualization of Project AHTAP—a space business center designed to safeguard our planet.

Project AHTAP, integrated with over 200 other projects, aims to generate sufficient tax revenue to fund all space programs initiated by Mr. Georgiy S. Garbuz. To materialize the construction of the Space Business Center, the following steps are required:

1. **Financing:** Secure the necessary funding.

2. **Selection of Professional Employees and Specialists:** Assemble a team of experts.
3. **Selection and Purchase of Special Materials and Technologies:** Acquire resources for design and construction.
4. **Creation of Space Business Center Design:** Develop a comprehensive design plan.
5. **Building of Space Business Center:** Execute the construction phase.

The revenue generated from the sales of the Space Business Center will be allocated to a space project program led by Georgiy S. Garbuz. This program aims to establish the National Garbuz Space Academy and affiliated schools worldwide, offering grants to foster care and underprivileged children, ensuring access to education.

The creation of Project AHTAP not only marks a significant step in advancing space exploration but also contributes to the broader goal of fostering education and scientific development on a global scale.

The project was initially conceived by Georgiy Sergeyevich Garbuz in 2007 and is being developed under the name "Project For The Future" (G.S. Garbuz) at the National Garbuz Space Academy Trust Corporation

---

PROJECT SEREGA

SEREGA, the International Space Government Grant Program, a groundbreaking initiative designed for outstanding Grade A (5) or B (4) students who excel in subjects like physics, mathematics, biology, fitness, and other school programs. Developed as part of the international space program in Garbuz Space School Academies and Garbuz Space Academies, SEREGA opens doors for exceptional students worldwide.

Eligibility for the SEREGA grant program is granted to students who achieve excellent grades at the completion of the 8th grade and those finishing high school at the 11th grade. Upon meeting the criteria, students gain the opportunity to enroll in the SEREGA grant program, an essential step towards education in space projects. Georgiy Garbuz's space program enrolls 30 thousand academy students in a country, totaling 8.6 million academy students globally.

Funding for student grants is derived from directed taxes associated with Georgiy Garbuz's scientific projects, works, and businesses. The program is named after a visionary individual who significantly contributed to the Soviet Union and later pioneered the introduction of German cars into the country. His communication skills and connections facilitated the opening of businesses with international companies, aiding millions in establishing enterprises and enhancing delivery services.

This influential figure also recognized the importance of education and personally engaged in teaching, providing invaluable skills to thousands of students. His commitment to education has resulted in a million exceptionally educated individuals prepared to contribute innovative ideas to society.

To bring the SEREGA program to fruition, the following steps are crucial:

1. Financing
2. Selection of professional employees and specialists
3. Selection and purchase of special materials and technologies for development
4. Opening government facilities for public access

Sales revenue generated will be directed to the Space Project Program initiated by Georgiy S. Garbuz. This program aims to build national Garbuz Space Academies and National Garbuz Space Academy Schools worldwide, offering educational opportunities to foster care children and other deserving students.

The project was initially conceived by Georgiy Sergeyevich Garbuz in 2007 and is being developed under the name "Project For The Future" (G.S. Garbuz) at the National Garbuz Space Academy Trust Corporation

---

GARBUZ SPACE SCHOOL ACADEMIES

The Garbuz Space School Academies present an international-intergalactic military educational space program aimed at fostering the intellectual wealth of nations. Focused on preparing students from grade 8 with excellent grades in physics, mathematics, biology, health, and other disciplines, the program paves the way for a future in space-related professions such as space doctors, astronauts, space scientists, space engineers, and many more. These academies are designed to accommodate 30 thousand students in each country and contribute to the education of 5.9 million students globally.

The primary objective of the Garbuz Space School Academies is to empower students with excellent academic achievements, preparing them for further education in space academies to attain degrees with top grades. The overarching mission is threefold: to provide quality education to outstanding students, to spearhead research and colonization efforts in space, and to establish a conducive living environment for everyone on Earth.

To bring this vision to reality, certain key steps must be taken:

1. **Financing:** Adequate financial support is essential to establish and sustain the space school academies, ensuring access to cutting-edge technology and resources.
2. **Selection of Professional Employees and Specialists:** Employing a dedicated team of educators and specialists is critical to maintaining the high standards of education and research within the academies.

3. **Selection and Purchase of Special Materials, Technologies for Development:** Acquiring state-of-the-art materials and technologies is imperative to provide students with a comprehensive and advanced learning experience.

4. **Opening Government Facilities for Public Access:** Creating accessible government facilities will enable public engagement, fostering a sense of community involvement and support for the space education initiative.

Sales revenue generated by the Garbuz Space School Academies will be channeled into a space project program initiated by Georgiy S. Garbuz. This program aims to establish national Garbuz Space Academies and National Garbuz Space Academy Schools globally, providing educational opportunities for foster care children and other deserving students. Funding for grants to educate children will be sourced from directed taxes on Georgiy S. Garbuz's science works and all associated projects.

Garbuz Space School Academies stand at the forefront of an innovative educational approach, nurturing the minds of exceptional students and contributing to the exploration and understanding of the vast realms of space. The success of this program not only elevates the intellectual capabilities of individuals but also propels nations toward a brighter, more informed future.

The project was initially conceived by Georgiy Sergeyevich Garbuz in 2007 and is being developed under the name "Project For The Future" (G.S. Garbuz) at the National Garbuz Space Academy Trust Corporation

---

FIRST STEP

The first step in the government's social program aims to inspire individuals to work, save money, and start businesses. The program offers consultations, develops personalized plans for participants, provides psychological support, and manages individuals' finances. It also assists in finding well-paying jobs and helps in initiating entrepreneurial ventures.

According to our research, individuals working at a Hyatt Hotel for $13 per hour are left with only $300 for living expenses after paying taxes and other costs. These people urgently require government assistance to sustain their livelihoods.

For instance, working as a delivery driver at Domino's for $24 per hour in a full-time job allows individuals to save $2,000 a month, participating in a government program. Through social consultations provided by the program, individuals can save over three years to open a partner Domino's store in the United States, including a $75,000 down payment and a $20,000 initial payment.

Similarly, working in a waste management company for $23 per hour in a full-time job enables individuals to save $2,000 a month, offering an opportunity to save and invest in a future business.

The first step program helps people realize their dreams and plans while addressing health problems. To implement this social program, the government needs:

1. Financing

2. Selection of professional employees and specialists
3. Selection and purchase of special materials and technologies for development
4. Establishment of government facilities for public assistance

Revenue generated from the program's sales will be directed towards a space project program initiated by Georgiy S. Garbuz to build the National Garbuz Space Academy and National Garbuz Space Academy schools worldwide, providing education opportunities for foster care and other children.

The project was initially conceived by Georgiy Sergeyevich Garbuz in 2007 and is being developed under the name "Project For The Future" (G.S. Garbuz) at the National Garbuz Space Academy Trust Corporation

---

EAGLE COMPUTER PROGRAM

Development of an enhanced voice-conversational computer program designed for accelerated and improved language learning, targeting school academies and facilitating the education, training, and practice of foreign languages for orphaned children and individuals of all ages.

The EAGLE computer program was devised to offer a comprehensive language learning experience for children participating in projects like "Kids of the Future" and others. Our research highlighted the challenge people face when learning languages solely from books, word by word. As part of our study, we engaged a Russian immigrant with limited English proficiency who resided in a shelter with 150 American individuals of various backgrounds. Despite the initial difficulty, communication improved over time through practical conversations, gestures, and shared experiences. By the end of the month, the immigrant demonstrated excellent English language proficiency, emphasizing the efficacy of practical language conversations.

Believing in the effectiveness of practical talk and conversations, we developed the EAGLE computer program for children and individuals of all ages, contributing to space projects like "Kids of the Future" and others.

To make the EAGLE computer program publicly accessible, we require:

1. Financing
2. Selection and hiring of professional employees and specialists

3. Creation of an enhanced voice and conversational computer program
4. Negotiation of international contracts for the sale and distribution of the language learning program to children's homes and individuals of all ages.

Proceeds from the program's sales will be directed toward a space project initiated by Georgiy S. Garbuz to establish the National Garbuz Space Academy and National Garbuz Space Academy schools worldwide, providing education opportunities for foster care and other children.

The project was initially conceived by Georgiy Sergeyevich Garbuz in 2007 and is being developed under the name "Project For The Future" (G.S. Garbuz) at the National Garbuz Space Academy Trust Corporation

---

CENT BOOST

Introducing the "CENT BOOST" centralized automatic currency trading program, developed to stimulate the economy, enhance bank trading, bolster military resources, and promote international trade, among other objectives. This program is fully centralized by the government and central bank to ensure stability in financial markets, job markets, and more.

In our research, we observed that converting 1 USD to 1 EUR and then back to USD could yield a profit of 0.01 cent. For example, investing $25,000 in the program could result in various currency trades, as illustrated below:

Minimum Investment: $5,000 USD

Maximum Investment: $25,000 USD

Currency	Profit	Fee
Tenge	$150	$60
Euro	$150	$60
Ruble	$150	$60
US Dollar	$270	$81

The program automatically analyzes and trades currencies at regular prices, allowing banks to increase sales fees and enabling individuals to earn up to $1,000 a day. This approach contributes to economic growth, boosts bank profits, and enhances individual wealth. To ensure responsible usage, the program is restricted to individuals working regular jobs, government military agents, and those supporting the government. The program automatically verifies users' IRS accounts to confirm their full-time employment status.

The daily circulation supply of the program is $100 quadrillion, and the currency is traded as needed with government approval to provide economic assistance to countries.

To make the program available to the public, we require:

1. Financing
2. Selection of study premises
3. Selection and purchase of special technologies for study and development
4. Selection and hiring of professional staff

Proceeds from the program's sales will be directed toward a space project program initiated by Georgiy S. Garbuz to establish the National Garbuz Space Academy and National Garbuz Space Academy schools worldwide, providing education opportunities for foster care and other children.

The project was initially conceived by Georgiy Sergeyevich Garbuz in 2007 and is being developed under the name "Project For The Future" (G.S. Garbuz) at the National Garbuz Space Academy Trust Corporation

---

PROJECT OTECH

Development of an Electric Gun for the Neutralization of
Nuclear Combat Atomic Missiles

In response to the escalating threat of warfare and the potential destruction posed by nuclear combat atomic missiles, we are initiating the creation of an electric gun designed to enhance security, safeguard borders, and protect states.

The current global landscape is fraught with dangers, with countries equipped with an array of weapons capable of causing massive destruction, including missiles, tanks, warships, and more. Our focus is on developing an electric gun as a proactive measure to secure borders. Our research indicates that a direct hit from an electric gun delivering 1000 volts can incapacitate tanks and missiles by disrupting their systems. A 10,000-volt hit can deactivate nuclear missiles, warships, tanks, and other military technology. A staggering 100,000-volt hit can utterly obliterate missiles, tanks, and warships.

By advancing the development of the electric gun, we aim to fortify our borders and prevent any external threats from compromising the safety of our cities and citizens.

To make the electric gun available for military use, we require:

1. Financing
2. Selection of professional scientific personnel

3. Selection of an Academy and facilities for scientific research and electric gun development
4. Electric gun design
5. Technology research and development
6. Selection of appropriate technologies and materials
7. Electric gun creation
8. Testing the electric gun at a designated test site

Proceeds from the sales of the electric gun will be allocated to a space project program initiated by Georgiy S. Garbuz, aimed at establishing the National Garbuz Space Academy and National Garbuz Space Academy schools worldwide, providing education opportunities for foster care and other children.

The project was initially conceived by Georgiy Sergeyevich Garbuz in 2007 and is being developed under the name "Project For The Future" (G.S. Garbuz) at the National Garbuz Space Academy Trust Corporation

---

10 YEARS OF GOOD LIFE

Government Medical Program for the Treatment of Depression and Other Diseases

---

The government is launching a medical program aimed at treating depression and various diseases, with the goal of enabling people to lead fulfilling lives. According to our research, a significant portion, approximately 70% of the global population, suffers from depression. The critical issue is the lack of guidance on how to spend their time, leading to thoughts of suicide or aggression. This is particularly prevalent among the unemployed and older individuals. In response, we have created the 10 Years Of Good Life Medical Program.

The program mandates that participants secure full-time employment, work five days a week, visit parks, restaurants, and engage in creative activities such as painting, writing books, composing songs, or storytelling. Regular doctor visits and adherence to a structured daily routine help individuals stay busy with exciting activities, eliminating thoughts of suicide or aggression. The medical procedure will cost $4000 per month for each patient, and our program aims to completely alleviate depression, enabling participants to live fulfilling and enjoyable lives.

To implement this medical treatment program for the public, we need:

1. Financing
2. Revenue from taxes
3. Funding allocation for public health programs

The revenue generated from this program will be directed towards a space project initiated by Georgiy S. Garbuz, intended to establish the National Garbuz Space Academy and National Garbuz Space Academy schools worldwide, providing educational opportunities for foster care and other children.

The project was initially conceived by Georgiy Sergeyevich Garbuz in 2007 and is being developed under the name "Project For The Future" (G.S. Garbuz) at the National Garbuz Space Academy Trust Corporation

---

MEDICAL TREATMENT OF CHILDREN AND PEOPLE WITH DISABILITIES OF ANY AGE, SUCH AS INFANTILE PARALYSIS AND OTHER PARALYTIC DISEASES, FOR THE RAPID DEVELOPMENT AND RESTORATION OF THE WORKING CAPACITY OF THE WHOLE ORGANISM.

People and children with disabilities, such as those with infantile paralysis and other paralytic diseases, often face severe mental health challenges. Depression, thoughts of death, suicide, aggression, and rudeness are common, largely attributed to excessive free time. Our research indicates that individuals with disabilities, like those who use wheelchairs, often contemplate suicide as they imagine various scenarios, providing a disturbing sense of relief to their minds.

To address these challenges, our medical program for people with disabilities seeks to make their dreams come true. This includes engaging in exercises to improve mobility, facilitating exciting conversations with healthcare professionals, and providing essential services to enhance their overall well-being.

To implement this medical treatment program for the public, we need:

1. Financing

2. Selection of premises for study
3. Selection and purchase of special technologies for the study and creation of sports physical exercises
4. Selection of professional specialists
5. Hiring and training of nurses for mandatory patient supervision
6. Selection of patients for the study
7. Creation and development of medical drugs for sports physical exercises on patients
8. Testing medical drugs for potential side effects
9. Treating patients with paralysis

The revenue generated from this program will be directed towards a space project initiated by Georgiy S. Garbuz, intended to establish the National Garbuz Space Academy and National Garbuz Space Academy schools worldwide, providing educational opportunities for foster care and other children.

The project was initially conceived by Georgiy Sergeyevich Garbuz in 2007 and is being developed under the name "Project For The Future" (G.S. Garbuz) at the National Garbuz Space Academy Trust Corporation

---

KIDS

---

*Design and Development of the Simulator for Leg Rehabili-
tation for Home Use*

Health is paramount in human life. Unfortunately, many people, both kids and adults, experience paralysis, leading to a life of depression. Our research reveals that a significant cause of depression is the abundance of free time, often leading individuals to contemplate suicide or become aggressive. It is crucial for them to find engaging activities to fill their time, and our research shows that activities like painting, walking, exercising, and others contribute to a more fulfilling life. Paralyzed adults and kids, resembling children who have never walked, anticipate positive changes. To address this, we have developed an automatic exercise machine for paralyzed individuals of all ages, helping them initiate walking and steering clear of depression to lead better lives.

To make the exercise machine available for public use, we need:

1. Financing
2. Selection of professional employees and specialists
3. Selection and procurement of special materials and technologies for creating the simulator
4. Simulator design and development
5. Contracts with corporations for production

6. Contracts for the global sale of the simulator

The revenue generated from the sales of the simulator will be directed towards a space project program created by Georgiy S. Garbuz. This initiative aims to establish the National Garbuz Space Academy and National Garbuz Space Academy schools globally, providing educational opportunities for foster care and other children.

The project was initially conceived by Georgiy Sergeyevich Garbuz in 2007 and is being developed under the name "Project For The Future" (G.S. Garbuz) at the National Garbuz Space Academy Trust Corporation

---

PENICILLIN PRO

Treatment Breakthrough for Lung Diseases

This innovative method, developed through meticulous scientific work, shows a 99.9% positive result in early private medical research. The treatment, named "Penicilin Pro," involves inhaling 100mg of the medication into the right nostril and 100mg into the left nostril using a NAT inhaler. After 60 seconds, a quick nosebleed concludes the process. The medication, beginning to act within 6 seconds, provides numerous benefits, including improved body performance, lung cleansing, enhanced vein functionality, and a healthier respiratory system.

The estimated cost for a 2-month treatment is $1600, with projected annual sales of $1.588 trillion in the USA and $3.188 trillion worldwide. Tax revenues generated from sales will be directed to a space project program initiated by Georgiy S. Garbuz, aiming to establish the National Garbuz Space Academy and National Garbuz Space Academy schools globally, supporting the education of foster care children.

This medical breakthrough marks a historic moment in the industry, with plans to create numerous medications, including treatments for respiratory organs and lungs, addressing even the most dangerous diseases like lung cancer.

To bring this vision to reality, the following steps are crucial:

1. Financing
2. Selection of professional staff and specialists from the Academy of Sciences and medical professionals
3. Procurement of special materials and technologies for medical examination
4. Patient selection for studying the new treatment method
5. Examination of the dangers and side effects of penicillin inhalation drugs
6. Discovery and study of new diseases during the treatment process
7. Continuous monitoring for potential side effects
8. Ongoing development of new medical drugs
9. Treatment of patients with lung cancer

A mandatory ban on usage without a doctor's recommendation is emphasized, as incorrect dosing poses severe risks to human health.

The project was initially conceived by Georgiy Sergeyevich Garbuz in 2007 and is being developed under the name "Project For The Future" (G.S. Garbuz) at the National Garbuz Space Academy Trust Corporation

---

NAT INHALER

Design and Development of the NAT Inhaler Medical Device
for Body Cleansing and Disease Treatment

The lungs, a vital organ in the human body, play a crucial role in ensuring the proper functioning of every organ by supplying oxygen. Unfortunately, billions of people face health issues each year due to clogged microveins and veins, resulting in insufficient oxygen delivery to organs. Georgiy S. Garbuz presents a groundbreaking medical solution through the NAT Inhaler, a medical device designed for inhaling medication directly into the lungs to have a targeted impact on various organs.

The NAT Inhaler is a 5-inch tubule equipped with a hole for inserting medication. It features a holder underneath for stability and a top-turn mechanism for even inhalation through the nose. The equipment includes separate 100mg and 50mg spoons for transferring medication from bottles to the NAT Inhaler for inhalation through the nose to the lungs.

Developed in compliance with all legal requirements to prevent illegal narcotic use, the cost of each disposable unit is $5, sold in boxes containing 62 pieces, providing a two-month medical treatment for $300.

Predicted Sales:

- USA: $90 billion

- Worldwide: $2.1 quadrillion

To make this medical equipment available to the public, the following steps are essential:

1. Financing
2. Selection of professional employees and specialists
3. Design creation for the medical product
4. Selection and purchase of special materials and technologies for product creation
5. Creation of the medical product
6. Patient selection for studying the created medical product
7. Creation and development of medical drugs related to the product
8. Checking medical drugs for side effects
9. Treatment of patients, body purification, lung cancer treatment, and other diseases

Revenues generated from sales will be directed to a space project program initiated by Georgiy S. Garbuz, aimed at establishing the National Garbuz Space Academy and National Garbuz Space Academy schools worldwide, supporting the education of foster care children.

The project was initially conceived by Georgiy Sergeyevich Garbuz in 2007 and is being developed under the name "Project For The Future" (G.S. Garbuz) at the National Garbuz Space Academy Trust Corporation

---

LUNGS ATTACK

Study of a New Lung Disease: "Lung Attack" from Penicillin and Narcotics Inhalation

---

The vulnerability of the lungs, as a crucial organ in the human body, has become apparent, especially with diseases like COVID-19 spreading through the air via breathing. In our research on medication inhalation into the lungs, we have identified a potential side effect, a new disease named "LUNGS ATTACK." Our findings indicate that individuals exceeding prescribed limits of medication or using narcotics experience lung blockage after just 5 minutes of use. This results in the lungs being unable to recycle the medication promptly, leading to a chemical reaction affecting microveins in the lungs, heart, and other organs. Patients may start feeling discomfort, which escalates into pain, accelerated breathing, and a sensation of blocked lungs. Our research reveals that individuals who misuse narcotics can suffer from lung attacks, potentially resulting in fatal consequences in cases of overdose. Further research is crucial to finding appropriate treatments for affected patients.

Predicted Sales:

- USA: $121 billion
- Worldwide: $4.90 quadrillion

To explore this new disease, the following steps are essential:

1. Financing
2. Selection of professional staff, including specialists from the Academy of Sciences and doctors
3. Selection and purchase of special materials and technologies for medical examination
4. Selection of patients for the study of the new disease
5. Studying the danger and characteristics of the new human disease
6. Detection and study of new types of diseases in the treatment of this condition
7. Checking medical drugs for potential side effects
8. Creation and development of new medical drugs during the study of the disease
9. Treatment of patients

Revenues from sales will be directed towards a space project program initiated by Georgiy S. Garbuz, aimed at establishing the National Garbuz Space Academy and National Garbuz Space Academy schools worldwide, supporting the education of foster care children.

A mandatory ban on use without a doctor's recommendation and prescription is crucial, as incorrect dosing of penicillin and narcotics can be extremely dangerous for human health.

The project was initially conceived by Georgiy Sergeyevich Garbuz in 2007 and is being developed under the name "Project For The Future" (G.S. Garbuz) at the National Garbuz Space Academy Trust Corporation

---

RESTORATION OF BRAIN FUNCTION IN THE HUMAN BODY

The brain is the most important organ in the human body, responsible for the accurate functioning of all bodily systems. However, every day, thousands of patients, often from car accidents, wars, and other incidents, are transported to hospitals with brain damage that could lead to lasting disabilities. The medical community faces challenges in providing effective treatment, prompting this scientific initiative by Georgiy S. Garbuz, aimed at researching and training medical professionals to offer improved care for such patients.

To bring this medical treatment to the public, the following steps are necessary:

1. **Financing:** Secure funding for extensive research and implementation.
2. **Laboratory Selection:** Choose a state-of-the-art laboratory for the theoretical study of the human body and the skull.
3. **Professional Staff:** Assemble a skilled and dedicated team to study the human body.
4. **Patient Selection:** Identify suitable patients (aged 75-90) for the theoretical study of the organism.
5. **Material Procurement:** Purchase necessary materials and ingredients for the research.
6. **Method Expansion:** Develop an extended method for studying the organism, brain functions, and their

compatibility for duration and healing in the human body.

7. **Brain Study:** Expand the method for studying the human brain.
8. **Body Study:** Extend the method for studying the human body.
9. **Surgical Operation:** Conduct a surgical operation for the implantation of the brain into the human body and focus on rapid healing.
10. **Post-Surgery Study:** Develop an extended method for studying the human brain and body after a surgical operation.
11. **Intravenous Study:** Expand the method for studying intravenous brain functions and their healing after surgery.

It's crucial to note that the proposed treatment is strictly prohibited for practical surgery on individuals under 75-90 years of age. The method has not been tested on humans to date, and its use in neurosurgery remains unexplored.

Strict confidentiality is maintained, and revenue generated from sales will be directed toward a space project program initiated by Georgiy S. Garbuz. This program aims to establish national Garbuz Space Academy schools globally, supporting foster care and providing education for children.

The project was initially conceived by Georgiy Sergeyevich Garbuz in 2007 and is being developed under the name "Project For The Future" (G.S. Garbuz) at the National Garbuz Space Academy Trust Corporation

---

RESTORING THE VIABILITY OF THE HUMAN BODY AND NORMAL BRAIN FUNCTIONING AFTER SURGICAL BRAIN IMPLANTATION

The brain, the most essential organ in the human body, is responsible for orchestrating all bodily functions. However, brain damage poses significant challenges to leading a fulfilling life. The development of training programs for doctors in the surgical operation of brain implantation into the body represents a groundbreaking initiative. This program aims to offer effective medical treatment to patients suffering from brain damage diseases while enhancing the professionalism of doctors globally. Such surgeries, never attempted in the history of medical treatment, require extensive study and practice to assist those in need.

To establish this training program, we require:

1. **Funding:** Secure financial resources for the research, development, and implementation of the training.
2. **Professional Team:** Hire and assemble a team of skilled professionals dedicated to training doctors.
3. **Technological Infrastructure:** Select, purchase, and create cutting-edge technologies necessary for studying and conducting surgical operations.
4. **Research Facility:** Acquire a suitable building equipped for advanced scientific research.

5. **Patient Selection:** Identify appropriate patients for scientific research and practical training.
6. **Theoretical Study:** Conduct an extensive theoretical study of the human body.
7. **Brain Study:** Undertake an in-depth theoretical study of the human brain.

The details of this program are classified as top secret. The revenue generated from its activities will be directed towards a space project program initiated by Georgiy S. Garbuz. This project aims to establish national Garbuz Space Academies and schools worldwide, providing education opportunities for foster care children and others in need.

The project was initially conceived by Georgiy Sergeyevich Garbuz in 2007 and is being developed under the name "Project For The Future" (G.S. Garbuz) at the National Garbuz Space Academy Trust Corporation

---

BLOODSAP

Medical Product Equipment for Treatment and Ensuring
Maximum Vitamin Access

The human body is susceptible to various diseases, and our research has revealed a unique approach to addressing deficiencies. Through plasma transfusion, we observed a patient experiencing microvein clearance issues and a deficiency in essential vitamins, leading to an unusual craving for unconventional substances. In response to this, we have developed the medical equipment called BLOODSAP, designed to provide vital vitamins to the human body.

To make this medical treatment available to the public, we require:

1. **Funding:** Obtain financial support for research, development, and production.
2. **Testing and Approval:** Conduct thorough testing and seek regulatory approvals for the BLOODSAP equipment.
3. **Production Facilities:** Establish facilities for the production of BLOODSAP equipment.
4. **Distribution Network:** Develop a reliable distribution network to make the product accessible to the public.

Revenue generated from the sales of BLOODSAP will be directed toward a space project program initiated by Georgiy S. Garbuz. This project aims to build national Garbuz Space

Academies and schools worldwide, providing educational opportunities for foster care children and others in need.

The project was initially conceived by Georgiy Sergeyevich Garbuz in 2007 and is being developed under the name "Project For The Future" (G.S. Garbuz) at the National Garbuz Space Academy Trust Corporation

---

CREATION OF AN OINTMENT FOR THE TREATMENT OF MILD BRAIN DAMAGE.

The brain, the most vital organ in the human body, is responsible for the proper functioning of all bodily processes. In today's world, numerous diseases pose significant challenges to people's well-being. Fortunately, through the scientific advancements of Georgiy S. Garbuz, we can potentially assist billions of individuals globally through medical treatments.

In our recent research, we focused on individuals with traumatic brain injuries (TBI), commonly known as brain damage. We administered a medication called Penicillin Pro through the nasal route. Using specialized medical equipment, we introduced loose powder into the nasal cavity, allowing the medication to reach the brain's microveins. Within 5 minutes, patients reported a noticeable improvement in brain performance, stability, and breakthroughs in previously clogged microveins.

We envision developing a Penicillin Pro ointment to be applied to the nasal marrow for 30 minutes, potentially providing fast treatment within six months, depending on the specific disease. This ointment could also be applied to wounds for neutralizing harmful molecules and promoting rapid healing. The 150ml bottle is estimated to cost $75, with projected sales of $22.5 billion in the USA and $525 trillion worldwide.

To bring this medication to the public, we require:

1. Financing
2. Selection of a laboratory for drug creation
3. Hiring professional staff, including specialists from the Academy of Sciences and doctors
4. Selection and acquisition of materials and technologies for drug creation
5. Identifying patients for studying the new treatment method
6. Studying the potential dangers of the medication for mild brain damage
7. Detecting and researching new types of diseases treatable with the new method
8. Conducting thorough checks for potential side effects
9. Continuous creation and development of new medical drugs while studying the new treatment method
10. Treating patients with mild brain damage using the developed medical treatment method, with a mandatory ban on usage without doctor recommendation.

Misuse of drug dosage is extremely hazardous to human health; therefore, strict adherence to doctor recommendations is crucial.

Revenue generated from sales will be allocated to a space project program initiated by Georgiy S. Garbuz, aimed at establishing the National Garbuz Space Academy and related schools worldwide. The program's primary goal is to provide grants for foster care and underprivileged children to receive education.

The project was initially conceived by Georgiy Sergeyevich Garbuz in 2007 and is being developed under the name "Project For The Future" (G.S. Garbuz) at the National Garbuz Space Academy Trust Corporation

MEDICAL TREATMENT OF HUMAN POTENCY! METHOD OF PSYCHOLOGICAL INFLUENCE ON THE HUMAN BRAIN USING ARTWORK OF NUDE WOMEN AND MEN DRAWN BY CHILDREN AGED 2 TO 13.

Sex is a crucial element in the life of every family. In our research, we focused on a patient experiencing potency issues due to chronic prostatitis. Employing psychological treatment, we aimed to address the patient's essential thoughts related to arousal and the desire for sexual relations.

During this study, we utilized psychological interventions to stimulate the patient's sexual interest. We presented images depicting sexual acts drawn by children aged 2 to 13. Upon reviewing these drawings, the patient reported experiencing dreams involving children and expressed a desire to have a family. This realization motivated him to seek a romantic relationship with a woman to start a family.

The psychological treatment proved highly effective, considering that many individuals struggle to concentrate due to life challenges. People often find it difficult to simply live, create, and enjoy life, with their primary focus often revolving around financial concerns. However, the images created by children elicited wishes and provided a sense of relaxation, normalizing brain functions.

We believe that incorporating paintings created by children in schools could significantly contribute to the well-being of adults. Such artwork has the potential to improve health, foster care for partners and children, and offer psychological relief. To implement this medical approach on a broader scale, we require:

1. An expanded methodology for studying and influencing the brain.
2. Enhanced artistic endeavors to facilitate an extensive exploration of the brain.

The project was initially conceived by Georgiy Sergeyevich Garbuz in 2007 and is being developed under the name "Project For The Future" (G.S. Garbuz) at the National Garbuz Space Academy Trust Corporation

---

MEDICAL TREATMENT OF THE RESPIRATORY ORGANS, LUNGS OF A HUMAN FROM LUNG DIASISES, USING THE INHALATION OF MEDICATIONS INTO THE LUNGS.

The lungs, a critical and vulnerable organ in the human body, face a myriad of diseases each year, presenting substantial challenges for both patients and medical professionals. Conditions ranging from Adjustment Disorder to Pulmonary Diseases and beyond necessitate constant medical attention. Despite the tireless efforts of doctors and an annual allocation of trillions of dollars for research, respiratory diseases persist as a formidable problem, claiming countless lives and causing immeasurable pain.

Georgiy S Garbuz introduces a groundbreaking medical treatment for respiratory diseases, specifically targeting the lungs. Through the inhalation of medications directly into the lungs, this innovative method promises a transformative approach to treating lung ailments. With initial research yielding a remarkable 99.9% success rate, the treatment offers a range of benefits, including improved overall body performance, lung cleansing, enhanced vein function, and optimized respiratory system performance.

The process involves using a specially formulated medication named "Penicillin Pro" through a NAT inhaler. Inhaling 100mg into each nostril, followed by a brief period, initiates a quick and effective treatment lasting for an entire day. The anticipated cost for a two-month treatment course is estimated at $1600.

Projected sales indicate a significant impact, with an estimated $1.59 trillion in the USA and $3.12 trillion globally. Tax revenues from these sales will be directed towards a space project program envisioned by Georgiy S Garbuz, aiming to establish national Garbuz space academies and schools worldwide, benefiting foster care children.

To achieve this medical breakthrough and further developments, the following steps are crucial:

1. Financing
2. Selection of professional staff, including specialists from the Academy of Sciences and doctors
3. Acquisition of special materials and technologies for medical examination
4. Identification of patients for studying the new treatment method
5. Examination of potential risks associated with penicillin inhalation for treating respiratory organs
6. Identification and study of new diseases related to the treatment method
7. Rigorous testing of medical drugs for potential side effects
8. Ongoing creation and development of new medical drugs
9. Treatment of patients suffering from lung cancer using the newly developed method, with a mandatory ban on unauthorized usage without a doctor's recommendation to avoid potential health risks.

This historic moment in the medical industry opens the door to numerous medical breakthroughs, particularly in treating respiratory diseases and lung cancer. The proposed steps lay the foundation for future advancements, offering hope to countless individuals suffering from these debilitating conditions.

The project was initially conceived by Georgiy Sergeyevich Garbuz in 2007 and is being developed under the name "Project For The Future" (G.S. Garbuz) at the National Garbuz Space Academy Trust Corporation

---

SEXUAL DEPRESSION

Depression has long been a pervasive issue in humanity, witnessed through centuries of suffering, suicides, assaults, and psychiatric hospitalizations. In contemporary times, our focus has shifted to a specific aspect known as sexual depression— an emerging diagnosis highlighting the widespread occurrence of sexual assaults worldwide. Millions of victims suffer daily, and many assailants go unidentified, perpetuating a cycle of crime and trauma.

In our research, we discovered a concerning correlation: a substantial number of prisoners charged with sexual assault had experienced a significant lack of sexual activity leading up to the crime. This raises questions about the role of sexual frustration in such offenses. To explore this, we conducted an experiment with a man accustomed to a regular sexual life involving prostitution once a week. After two weeks without sexual activity, he displayed escalating signs of frustration and aggression. By the fifth week, he had become a potential danger to society, planning sexual assaults.

This study underscores the impact of unmet sexual needs on an individual's mental state and behavioral tendencies. Recognizing the potential danger arising from sexual deprivation, we propose a comprehensive medical treatment plan to address sexual depression. To implement this on a public scale, we require:

1. Financing
2. Selection of professional medical staff

3. Inclusion of academic experts in sciences
4. Selection of appropriate patients for the study
5. Purchase of necessary materials for medicine creation and procedural development
6. Establishment of a medical treatment procedure
7. In-depth study of side effects during the creation of medical treatments
8. Male and female body reassurance treatments
9. Establishment of contracts for sexual services for patients

The revenue generated from these medical treatments will be allocated to a space project program initiated by Georgiy S. Garbuz. This program aims to build the National Garbuz Space Academy and affiliated schools worldwide, providing grants for foster care and underprivileged children to receive education.

Our commitment is not only to address pressing medical issues but also to contribute to the educational advancement and well-being of future generations on a global scale.

The project was initially conceived by Georgiy Sergeyevich Garbuz in 2007 and is being developed under the name "Project For The Future" (G.S. Garbuz) at the National Garbuz Space Academy Trust Corporation

---

LAZ DEPRESSION

Depression stands as one of the most pervasive challenges faced by humanity today, affecting nearly 65% of individuals across all walks of life, including the young, elderly, homeless, and many others. The consequences are far-reaching, manifesting in mass shootings perpetrated by those grappling with the weight of depression. As a witness to such incidents, I've observed the tragic outcomes of untreated mental health issues, shedding light on the urgent need for effective treatments.

In my 12-year journey battling depression, I have delved into extensive research to understand the underlying causes and potential solutions. The crux of my findings emphasizes the role of time – a crucial factor in the onset and progression of depression. A poignant example is the phenomenon of "Laz Depression," which often surfaces after significant life changes, such as retirement or job loss.

During my own struggle, I realized that actively filling my time with purposeful activities was a key element in alleviating depressive symptoms. Smoking served as a temporary distraction, offering a brief respite from intrusive thoughts. Engaging in artistic pursuits, such as drawing and painting, further extended my focus, providing relief for days on end. The pursuit of creative outlets acted as a vital lifeline, offering a reprieve from suicidal ideation.

Recognizing the importance of communication, I sought solace in conversations with my parents. While professional medical support is paramount, sharing the intricacies of life with loved ones played a crucial role in restoring my mental stability. The subsequent endeavor of writing a book became a monumental undertaking, consuming 150 days of my time and affording me moments of respite from the grips of depression.

In light of these personal discoveries, I propose a revolutionary approach to treating depression that emphasizes planned, engaging activities tailored to individual interests. This cost-effective method hinges on the concept of redirecting one's focus towards meaningful pursuits. To implement this approach on a public scale, the following components are required:

1. Financing
2. Selection of professional medical staff
3. Establishment of premises and laboratories for scientific research
4. Recruitment of patients for scientific studies
5. In-depth exploration of Laz Depression

Revenue generated from the proposed treatments will be channeled into a space project program envisioned by Georgiy S. Garbuz. This initiative aims to establish the National Garbuz Space Academy and affiliated schools worldwide, providing educational opportunities for foster care and underprivileged children.

Through sharing my personal journey, I hope to contribute to a paradigm shift in depression treatment, fostering a future where individuals can overcome this pervasive mental health challenge and find fulfillment in their lives.

The project was initially conceived by Georgiy Sergeyevich Garbuz in 2007 and is being developed under the name "Project For The Future" (G.S. Garbuz) at the National Garbuz Space Academy Trust Corporation

---

THE PSYCHOLOGICAL IMPACT OF DRAWING ARTWORKS ON THE RELAXATION OF THE HUMAN BRAIN.

The human brain, being the most important organ in the body, influences various aspects of our perception, including vision, hearing, feeling, smell, and more. Our research delves into the realm of depression, a pervasive issue affecting 70% of the global population. This study aims to understand the impact of time on the brain and its role in depression, exploring innovative ways to provide effective medical treatment.

Our findings reveal a profound connection between time management and mental health. In instances where individuals experience extended periods of unemployment or idleness, depression becomes a prevalent concern. To illustrate, we examined a patient who had been unemployed for five months, revealing that engaging in activities such as smoking temporarily diverted his thoughts from suicidal tendencies. However, once he ceased these activities, negative thoughts resurfaced, emphasizing the critical role of time in mental well-being.

Understanding the need for constructive engagement, we advocate for art as a therapeutic medium. Artistic endeavors, including painting and drawing, offer individuals an outlet for expression and creativity, occupying their minds for extended periods. This proves to be a cost-effective and powerful method to counteract suicidal thoughts and combat depression, with a small painting costing as little as $25-$30.

Projected costs for this medical treatment are estimated at $40 per day, with anticipated sales reaching $1.15 trillion in the USA and $15.9 trillion worldwide.

To enhance the quality of this medical treatment and further our understanding of brain diseases, we propose the following steps:

1. Project financing
2. Expansion of the method for studying brain diseases and their influence on brain relaxation
3. Advanced methods for studying and influencing the brain, including the impact of art on relaxation
4. Increased creation of art for extensive brain study
5. Advanced research and creation of medicines for brain diseases
6. In-depth study of potential side effects in the creation of drugs and medicines

The tax revenue generated from sales will contribute to a space project program initiated by Georgiy S Garbuz, aiming to establish national Garbuz space academies and schools worldwide, providing education opportunities for foster care children and others.

This revolutionary approach to depression treatment emphasizes the transformative power of art and its potential to alleviate the mental health struggles faced by millions.

The project was initially conceived by Georgiy Sergeyevich Garbuz in 2007 and is being developed under the name "Project For The Future" (G.S. Garbuz) at the National Garbuz Space Academy Trust Corporation

---

PROJECT HATHOR

Project Hathor is an international initiative aimed at legalizing and regulating prostitution across countries to provide a legal framework for individuals to engage in sex work, ensuring the creation of a safe and regulated environment. Our research, conducted in Minneapolis, MN, revealed that over 20 million individuals are involved in illegal sex services in the United States, with prices ranging from $150 per hour to $400 per night.

In downtown Minneapolis, we found homeless individuals offering sex services at a significantly lower price range of $20 to $40. Our survey of 1,000 women revealed that 993 were willing to provide services for $150 per hour, with only 7 declining due to marriage commitments. Currently, the United States government loses substantial tax revenue, estimated at $2.20 trillion annually, due to the illegal nature of prostitution. This money remains in the hands of criminals, contributing to various societal issues such as violence, shootings, and assaults.

Project Hathor proposes the legalization of prostitution globally, with the aim of redirecting the generated tax revenue towards healthcare, military agencies, education, and economic development. Legalizing this industry has the potential to provide free healthcare, quality education for excellent students, decent pay for military personnel, and specialized space education for children.

To implement Project Hathor publicly, we acknowledge that some may find it unconventional, but we stand as:

1. Realists who have witnessed the repercussions of the illegal sex trade.
2. Believers in the transformative power of redirecting funds for societal betterment.
3. Advocates for exploring and colonizing space for the prosperity of nations and the world.

For the successful development and execution of Project Hathor, we require:

1. Financing
2. Selection of professional employees and specialists
3. Selection and purchase of special materials and technologies for development
4. Opening government facilities for public transparency

Revenue generated from Project Hathor sales will be allocated to healthcare, military agency expenses, and the space project program initiated by Georgiy S. Garbuz. This program aims to establish the National Garbuz Space Academy and affiliated schools globally, providing grants to foster care and underprivileged children for their education. Funding for these grants will be sourced from directed taxes on Georgiy S. Garbuz's science works and all associated projects.

The project was initially conceived by Georgiy Sergeyevich Garbuz in 2007 and is being developed under the name "Project For The Future" (G.S. Garbuz) at the National Garbuz Space Academy Trust Corporation

---

HATHOR INTERNATIONAL SPACE BROTHEL STATE CORPORATION OF MINNESOTA

The International Intergalactic State Corporation is established with the dual objective of enhancing the economy at the national, state, and global levels, and exploring space across galaxies. With the support of the president, our initiative aims to legalize prostitution in the United States, with the aspiration to extend this legalization globally. Our plan involves providing sex services through designated brothels in various countries.

This endeavor addresses a lucrative yet illegal industry thriving in the black market. According to our research, a single prostitute serves 3-4 clients daily, contributing to a staggering annual profit of $2.20 trillion in the United States alone, with over 20 million individuals engaged in this profession. To fulfill the needs of the state of Minnesota, we propose the establishment of four international state corporations, generating a profit of $877 billion. Similarly, the state of New York would require eight corporations, and to cover the entire USA market, we estimate a need for 300 such entities.

Each state would control its prostitution market through a state corporation, allowing individuals to open partner brothels in every city. The transition from an illegal to a legal market is part of a new government project program aimed at legalizing former prostitutes and pimps. A special program has been devised, contingent on public acceptance, facilitating legal business operations and the provision of sex services within the country.

In summary, the International Intergalactic State Corporation envisions:

1. Legalizing and regulating the prostitution industry for economic growth.
2. Exploring space around galaxies for scientific advancements.
3. Implementing a government project program to transition from illegal to legal prostitution.
4. Establishing state-controlled corporations to oversee and regulate the legalized industry.

The revenue generated from these legalized operations will be instrumental in fulfilling various state and national needs, contributing to healthcare, education, military agencies, and space exploration. We believe that this innovative approach can not only boost the economy but also pave the way for progressive societal changes.

Hathor International Space Brothel State Corporation of Minnesota

- 40%: State Government
- 20%: Imperial Space Federation (USA Government)
- 10%: Federal Government
- 10%: NASA
- 5%: CIA, Interpol, FBI, or Military
- 5%: State Police or Federal Police
- 5%: Russia, Moldova, Mexico, or any other country government
- 5%: Local Businessmen, or Former Pimp or Prostitute (subject to meeting all government agency requirements)

Hathor International Space Brothel State Corporation of Ada, MN

- 40%: Hathor International Space Brothel State Corporation of Minnesota (Owner)
- 10%: Imperial Space Federation
- 50%: Local Businessmen or Former Pimp or Prostitute (subject to meeting all government agency requirements)

Hathor International Space Brothel State Corporation of Adams, MN

- 40%: Hathor International Space Brothel State Corporation of Minnesota (Owner)
- 10%: Imperial Space Federation
- 50%: Local Businessmen or Former Pimp or Prostitute (subject to meeting all government agency requirements)

Hathor International Space Brothel State Corporation of Adrian, MN

- 40%: Hathor International Space Brothel State Corporation of Minnesota (Owner)
- 10%: Imperial Space Federation
- 50%: Local Businessmen or Former Pimp or Prostitute (subject to meeting all government agency requirements)

Hathor International Space Brothel State Corporation of Afton, MN

- 40%: Hathor International Space Brothel State Corporation of Minnesota (Owner)
- 10%: Imperial Space Federation
- 50%: Local Businessmen or Former Pimp or Prostitute (subject to meeting all government agency requirements)

Hathor International Space Brothel State Corporation of Aitkin, MN

- 40%: Hathor International Space Brothel State Corporation of Minnesota (Owner)
- 10%: Imperial Space Federation
- 50%: Local Businessmen or Former Pimp or Prostitute (subject to meeting all government agency requirements)

Hathor International Space Brothel State Corporation of Akeley, MN

- 40%: Hathor International Space Brothel State Corporation of Minnesota (Owner)
- 10%: Imperial Space Federation
- 50%: Local Businessmen or Former Pimp or Prostitute (subject to meeting all government agency requirements)

Hathor International Space Brothel State Corporation of Albany, MN

- 40%: Hathor International Space Brothel State Corporation of Minnesota (Owner)
- 10%: Imperial Space Federation
- 50%: Local Businessmen or Former Pimp or Prostitute (subject to meeting all government agency requirements)

Hathor International Space Brothel State Corporation of Albert Lea, MN

- 40%: Hathor International Space Brothel State Corporation of Minnesota (Owner)
- 10%: Imperial Space Federation
- 50%: Local Businessmen or Former Pimp or Prostitute (subject to meeting all government agency requirements)

Hathor International Space Brothel State Corporation of Albertville, MN

- 40%: Hathor International Space Brothel State Corporation of Minnesota (Owner)
- 10%: Imperial Space Federation
- 50%: Local Businessmen or Former Pimp or Prostitute (subject to meeting all government agency requirements)

Hathor International Space Brothel State Corporation of Alden, MN

- 40%: Hathor International Space Brothel State Corporation of Minnesota (Owner)
- 10%: Imperial Space Federation
- 50%: Local Businessmen or Former Pimp or Prostitute (subject to meeting all government agency requirements)

Hathor International Space Brothel State Corporation of Alexandria, MN

- 40%: Hathor International Space Brothel State Corporation of Minnesota (Owner)
- 10%: Imperial Space Federation
- 50%: Local Businessmen or Former Pimp or Prostitute (subject to meeting all government agency requirements)

Hathor International Space Brothel State Corporation of Aitura, MN

- 40%: Hathor International Space Brothel State Corporation of Minnesota (Owner)
- 10%: Imperial Space Federation
- 50%: Local Businessmen or Former Pimp or Prostitute (subject to meeting all government agency requirements)

Hathor International Space Brothel State Corporation of Alvarado, MN

- 40%: Hathor International Space Brothel State Corporation of Minnesota (Owner)
- 10%: Imperial Space Federation
- 50%: Local Businessmen or Former Pimp or Prostitute (subject to meeting all government agency requirements)

Hathor International Space Brothel State Corporation of Amboy, MN

- 40%: Hathor International Space Brothel State Corporation of Minnesota (Owner)
- 10%: Imperial Space Federation
- 50%: Local Businessmen or Former Pimp or Prostitute (subject to meeting all government agency requirements)

Hathor International Space Brothel State Corporation of Andover, MN

- 40%: Hathor International Space Brothel State Corporation of Minnesota (Owner)
- 10%: Imperial Space Federation
- 50%: Local Businessmen or Former Pimp or Prostitute (subject to meeting all government agency requirements)

Hathor International Space Brothel State Corporation of Annandale, MN

- 40%: Hathor International Space Brothel State Corporation of Minnesota (Owner)
- 10%: Imperial Space Federation
- 50%: Local Businessmen or Former Pimp or Prostitute (subject to meeting all government agency requirements)

Hathor International Space Brothel State Corporation of Anoka, MN

- 40%: Hathor International Space Brothel State Corporation of Minnesota (Owner)
- 10%: Imperial Space Federation
- 50%: Local Businessmen or Former Pimp or Prostitute (subject to meeting all government agency requirements)

Hathor International Space Brothel State Corporation of Appleton, MN

- 40%: Hathor International Space Brothel State Corporation of Minnesota (Owner)
- 10%: Imperial Space Federation
- 50%: Local Businessmen or Former Pimp or Prostitute (subject to meeting all government agency requirements)

Hathor International Space Brothel State Corporation of Argyle, MN

- 40%: Hathor International Space Brothel State Corporation of Minnesota (Owner)
- 10%: Imperial Space Federation
- 50%: Local Businessmen or Former Pimp or Prostitute (subject to meeting all government agency requirements)

Hathor International Space Brothel State Corporation of Arlington, MN

- 40%: Hathor International Space Brothel State Corporation of Minnesota (Owner)
- 10%: Imperial Space Federation
- 50%: Local Businessmen or Former Pimp or Prostitute (subject to meeting all government agency requirements)

Hathor International Space Brothel State Corporation of Ashby, MN

- 40%: Hathor International Space Brothel State Corporation of Minnesota (Owner)
- 10%: Imperial Space Federation
- 50%: Local Businessmen or Former Pimp or Prostitute (subject to meeting all government agency requirements)

Hathor International Space Brothel State Corporation of Askov, MN

- 40%: Hathor International Space Brothel State Corporation of Minnesota (Owner)
- 10%: Imperial Space Federation
- 50%: Local Businessmen or Former Pimp or Prostitute (subject to meeting all government agency requirements)

Hathor International Space Brothel State Corporation of Atwater, MN

- 40%: Hathor International Space Brothel State Corporation of Minnesota (Owner)
- 10%: Imperial Space Federation
- 50%: Local Businessmen or Former Pimp or Prostitute (subject to meeting all government agency requirements)

In addition, Hathor International Space Brothel State Corporations will be created for the other following cities in the state of Minnesota, operating with the same ownership structure as above:

- Hathor International Space Brothel State Corporation of Audubon, MN
- Hathor International Space Brothel State Corporation of Aurora, MN
- Hathor International Space Brothel State Corporation of Austin, MN
- Hathor International Space Brothel State Corporation of Avon, MN
- Hathor International Space Brothel State Corporation of Babbitt, MN

- Hathor International Space Brothel State Corporation of Backus, MN
- Hathor International Space Brothel State Corporation of Badger, MN
- Hathor International Space Brothel State Corporation of Bagley, MN
- Hathor International Space Brothel State Corporation of Balaton, MN
- Hathor International Space Brothel State Corporation of Barnesville, MN
- Hathor International Space Brothel State Corporation of Barnum, MN
- Hathor International Space Brothel State Corporation of Barrett, MN
- Hathor International Space Brothel State Corporation of Battle Lake, MN
- Hathor International Space Brothel State Corporation of Baudette, MN
- Hathor International Space Brothel State Corporation of Baxter, MN
- Hathor International Space Brothel State Corporation of Bayport, MN
- Hathor International Space Brothel State Corporation of Beardlsey, MN
- Hathor International Space Brothel State Corporation of Beaver Creek, MN
- Hathor International Space Brothel State Corporation of Becker, MN
- Hathor International Space Brothel State Corporation of Belgrade, MN
- Hathor International Space Brothel State Corporation of Belle Plaine, MN
- Hathor International Space Brothel State Corporation of Bellingham, MN
- Hathor International Space Brothel State Corporation of Belview, MN
- Hathor International Space Brothel State Corporation of Bemidji, MN
- Hathor International Space Brothel State Corporation of Benson, MN
- Hathor International Space Brothel State Corporation of Bertha, MN
- Hathor International Space Brothel State Corporation of Bethel, MN
- Hathor International Space Brothel State Corporation of Big Falls, MN
- Hathor International Space Brothel State Corporation of Big Lake, MN
- Hathor International Space Brothel State Corporation of Bigelow, MN
- Hathor International Space Brothel State Corporation of Bigfork, MN
- Hathor International Space Brothel State Corporation of Bird Island, MN
- Hathor International Space Brothel State Corporation of Biwabik, MN
- Hathor International Space Brothel State Corporation of Blackduck, MN
- Hathor International Space Brothel State Corporation of Blooming Prairie, MN
- Hathor International Space Brothel State Corporation of Blue Earth, MN
- Hathor International Space Brothel State Corporation of Bluffton, MN
- Hathor International Space Brothel State Corporation of Bovey, MN

- Hathor International Space Brothel State Corporation of Bowlus, MN
- Hathor International Space Brothel State Corporation of Boyd, MN
- Hathor International Space Brothel State Corporation of Braham, MN
- Hathor International Space Brothel State Corporation of Brainerd, MN
- Hathor International Space Brothel State Corporation of Brandon, MN
- Hathor International Space Brothel State Corporation of Breckenridge, MN
- Hathor International Space Brothel State Corporation of Brewster, MN
- Hathor International Space Brothel State Corporation of Bricelyn, MN
- Hathor International Space Brothel State Corporation of Brooten, MN
- Hathor International Space Brothel State Corporation of Browerville, MN
- Hathor International Space Brothel State Corporation of Browns Valley, MN
- Hathor International Space Brothel State Corporation of Brownsdale, MN
- Hathor International Space Brothel State Corporation of Brownsville, MN
- Hathor International Space Brothel State Corporation of Brownton, MN
- Hathor International Space Brothel State Corporation of Buckman, MN
- Hathor International Space Brothel State Corporation of Buffalo Lake, MN
- Hathor International Space Brothel State Corporation of Buffalo, MN
- Hathor International Space Brothel State Corporation of Buhl, MN
- Hathor International Space Brothel State Corporation of Burnsville, MN
- Hathor International Space Brothel State Corporation of Butterfield, MN
- Hathor International Space Brothel State Corporation of Byron, MN
- Hathor International Space Brothel State Corporation of Caledonia, MN
- Hathor International Space Brothel State Corporation of Calumet, MN
- Hathor International Space Brothel State Corporation of Cambridge, MN
- Hathor International Space Brothel State Corporation of Campbell, MN
- Hathor International Space Brothel State Corporation of Canby, MN
- Hathor International Space Brothel State Corporation of Cannon Falls, MN
- Hathor International Space Brothel State Corporation of Canton, MN
- Hathor International Space Brothel State Corporation of Carlos, MN
- Hathor International Space Brothel State Corporation of Carlton, MN
- Hathor International Space Brothel State Corporation of Carver, MN
- Hathor International Space Brothel State Corporation of Cass Lake, MN
- Hathor International Space Brothel State Corporation of Center City, MN

- Hathor International Space Brothel State Corporation of Ceylon, MN
- Hathor International Space Brothel State Corporation of Champlin, MN
- Hathor International Space Brothel State Corporation of Chandler, MN
- Hathor International Space Brothel State Corporation of Chanhassen, MN
- Hathor International Space Brothel State Corporation of Chaska, MN
- Hathor International Space Brothel State Corporation of Chatfield, MN
- Hathor International Space Brothel State Corporation of Chisago City, MN
- Hathor International Space Brothel State Corporation of Chisholm, MN
- Hathor International Space Brothel State Corporation of Chokio, MN
- Hathor International Space Brothel State Corporation of Circle Pines, MN
- Hathor International Space Brothel State Corporation of Clara City, MN
- Hathor International Space Brothel State Corporation of Claremont, MN
- Hathor International Space Brothel State Corporation of Clarissa, MN
- Hathor International Space Brothel State Corporation of Clarkfield, MN
- Hathor International Space Brothel State Corporation of Clarks Grove, MN
- Hathor International Space Brothel State Corporation of Clear Lake, MN
- Hathor International Space Brothel State Corporation of Clearbrook, MN
- Hathor International Space Brothel State Corporation of Clearwater, MN
- Hathor International Space Brothel State Corporation of Cleveland, MN
- Hathor International Space Brothel State Corporation of Climax, MN
- Hathor International Space Brothel State Corporation of Clinton, MN
- Hathor International Space Brothel State Corporation of Cloquet, MN
- Hathor International Space Brothel State Corporation of Cohasset, MN
- Hathor International Space Brothel State Corporation of Cokato, MN
- Hathor International Space Brothel State Corporation of Cold Spring, MN
- Hathor International Space Brothel State Corporation of Coleraine, MN
- Hathor International Space Brothel State Corporation of Cologne, MN
- Hathor International Space Brothel State Corporation of Comfrey, MN
- Hathor International Space Brothel State Corporation of Cook, MN
- Hathor International Space Brothel State Corporation of Cosmos, MN
- Hathor International Space Brothel State Corporation of Cottage Grove, MN
- Hathor International Space Brothel State Corporation of Cottonwood, MN
- Hathor International Space Brothel State Corporation of Courtland, MN

- Hathor International Space Brothel State Corporation of Crookston, MN
- Hathor International Space Brothel State Corporation of Crosby, MN
- Hathor International Space Brothel State Corporation of Crosslake, MN
- Hathor International Space Brothel State Corporation of Crystal Bay, MN
- Hathor International Space Brothel State Corporation of Currie, MN
- Hathor International Space Brothel State Corporation of Cyrus, MN
- Hathor International Space Brothel State Corporation of Dakota, MN
- Hathor International Space Brothel State Corporation of Dalton, MN
- Hathor International Space Brothel State Corporation of Danube, MN
- Hathor International Space Brothel State Corporation of Darwin, MN
- Hathor International Space Brothel State Corporation of Dassel, MN
- Hathor International Space Brothel State Corporation of Dawson, MN
- Hathor International Space Brothel State Corporation of Dayton, MN
- Hathor International Space Brothel State Corporation of Deer Creek, MN
- Hathor International Space Brothel State Corporation of Deer River, MN
- Hathor International Space Brothel State Corporation of Deerwood, MN
- Hathor International Space Brothel State Corporation of Delano, MN
- Hathor International Space Brothel State Corporation of Delavan, MN
- Hathor International Space Brothel State Corporation of Detroit Lakes, MN
- Hathor International Space Brothel State Corporation of Dexter, MN
- Hathor International Space Brothel State Corporation of Dilworth, MN
- Hathor International Space Brothel State Corporation of Dodge Center, MN
- Hathor International Space Brothel State Corporation of Donnelly, MN
- Hathor International Space Brothel State Corporation of Dover, MN
- Hathor International Space Brothel State Corporation of Duluth, MN
- Hathor International Space Brothel State Corporation of Dundas, MN
- Hathor International Space Brothel State Corporation of Eagle Bend, MN
- Hathor International Space Brothel State Corporation of Eagle Lake, MN
- Hathor International Space Brothel State Corporation of East Grand Forks, MN
- Hathor International Space Brothel State Corporation of Easton, MN
- Hathor International Space Brothel State Corporation of Echo, MN
- Hathor International Space Brothel State Corporation of Eden Prairie, MN
- Hathor International Space Brothel State Corporation of Eden Valley, MN

- Hathor International Space Brothel State Corporation of Edgerton, MN
- Hathor International Space Brothel State Corporation of Eitzen, MN
- Hathor International Space Brothel State Corporation of Elbow Lake, MN
- Hathor International Space Brothel State Corporation of Elgin, MN
- Hathor International Space Brothel State Corporation of Elk River, MN
- Hathor International Space Brothel State Corporation of Elko, MN
- Hathor International Space Brothel State Corporation of Ellendale, MN
- Hathor International Space Brothel State Corporation of Ellsworth, MN
- Hathor International Space Brothel State Corporation of Elmore, MN
- Hathor International Space Brothel State Corporation of Ely, MN
- Hathor International Space Brothel State Corporation of Elysian, MN
- Hathor International Space Brothel State Corporation of Emily, MN
- Hathor International Space Brothel State Corporation of Emmons, MN
- Hathor International Space Brothel State Corporation of Erskine, MN
- Hathor International Space Brothel State Corporation of Evansville, MN
- Hathor International Space Brothel State Corporation of Eveleth, MN
- Hathor International Space Brothel State Corporation of Excelsior, MN
- Hathor International Space Brothel State Corporation of Eyota, MN
- Hathor International Space Brothel State Corporation of Fairfax, MN
- Hathor International Space Brothel State Corporation of Fairmont, MN
- Hathor International Space Brothel State Corporation of Faribault, MN
- Hathor International Space Brothel State Corporation of Farmington, MN
- Hathor International Space Brothel State Corporation of Felton, MN
- Hathor International Space Brothel State Corporation of Fergus Falls, MN
- Hathor International Space Brothel State Corporation of Fertile, MN
- Hathor International Space Brothel State Corporation of Fifty Lakes, MN
- Hathor International Space Brothel State Corporation of Finlayson, MN
- Hathor International Space Brothel State Corporation of Fisher, MN
- Hathor International Space Brothel State Corporation of Flensburg, MN
- Hathor International Space Brothel State Corporation of Floodwood, MN
- Hathor International Space Brothel State Corporation of Foley, MN
- Hathor International Space Brothel State Corporation of Forest Lake, MN
- Hathor International Space Brothel State Corporation of Foreston, MN

- Hathor International Space Brothel State Corporation of Fosston, MN
- Hathor International Space Brothel State Corporation of Fountain, MN
- Hathor International Space Brothel State Corporation of Foreston, MN
- Hathor International Space Brothel State Corporation of Fosston, MN
- Hathor International Space Brothel State Corporation of Fountain, MN
- Hathor International Space Brothel State Corporation of Franklin, MN
- Hathor International Space Brothel State Corporation of Frazee, MN
- Hathor International Space Brothel State Corporation of Freeborn, MN
- Hathor International Space Brothel State Corporation of Freeport, MN
- Hathor International Space Brothel State Corporation of Frost, MN
- Hathor International Space Brothel State Corporation of Fulda, MN
- Hathor International Space Brothel State Corporation of Garfield, MN
- Hathor International Space Brothel State Corporation of Garrison, MN
- Hathor International Space Brothel State Corporation of Gary, MN
- Hathor International Space Brothel State Corporation of Gaylor, MN
- Hathor International Space Brothel State Corporation of Geneva, MN
- Hathor International Space Brothel State Corporation of Ghent, MN
- Hathor International Space Brothel State Corporation of Gibbon, MN
- Hathor International Space Brothel State Corporation of Gilbert, MN
- Hathor International Space Brothel State Corporation of Gilman, MN
- Hathor International Space Brothel State Corporation of Glencoe, MN
- Hathor International Space Brothel State Corporation of Glenville, MN
- Hathor International Space Brothel State Corporation of Glenwood, MN
- Hathor International Space Brothel State Corporation of Glyndon, MN
- Hathor International Space Brothel State Corporation of Gonvick, MN
- Hathor International Space Brothel State Corporation of Good Thunder, MN
- Hathor International Space Brothel State Corporation of Goodhue, MN
- Hathor International Space Brothel State Corporation of Graceville, MN
- Hathor International Space Brothel State Corporation of Granada, MN
- Hathor International Space Brothel State Corporation of Grand Marais, MN
- Hathor International Space Brothel State Corporation of Grand Meadow, MN
- Hathor International Space Brothel State Corporation of Grand Rapids, MN
- Hathor International Space Brothel State Corporation of Granite Falls, MN

- Hathor International Space Brothel State Corporation of Green Isle, MN
- Hathor International Space Brothel State Corporation of Greenbush, MN
- Hathor International Space Brothel State Corporation of Greenwald, MN
- Hathor International Space Brothel State Corporation of Grey Eagle, MN
- Hathor International Space Brothel State Corporation of Grove City, MN
- Hathor International Space Brothel State Corporation of Grygla, MN
- Hathor International Space Brothel State Corporation of Hackensack, MN
- Hathor International Space Brothel State Corporation of Hallock, MN
- Hathor International Space Brothel State Corporation of Halstad, MN
- Hathor International Space Brothel State Corporation of Hamburg, MN
- Hathor International Space Brothel State Corporation of Hampton, MN
- Hathor International Space Brothel State Corporation of Hancock, MN
- Hathor International Space Brothel State Corporation of Hancock, MN
- Hathor International Space Brothel State Corporation of Hanley Falls, MN
- Hathor International Space Brothel State Corporation of Hanover, MN
- Hathor International Space Brothel State Corporation of Hanska, MN
- Hathor International Space Brothel State Corporation of Hardwick, MN
- Hathor International Space Brothel State Corporation of Harmony, MN
- Hathor International Space Brothel State Corporation of Harris, MN
- Hathor International Space Brothel State Corporation of Hartland, MN
- Hathor International Space Brothel State Corporation of Hastings, MN
- Hathor International Space Brothel State Corporation of Hawley, MN
- Hathor International Space Brothel State Corporation of Hayfield, MN
- Hathor International Space Brothel State Corporation of Hayward, MN
- Hathor International Space Brothel State Corporation of Hector, MN
- Hathor International Space Brothel State Corporation of Henderson, MN
- Hathor International Space Brothel State Corporation of Hendricks, MN
- Hathor International Space Brothel State Corporation of Hendrum, MN
- Hathor International Space Brothel State Corporation of Henning, MN
- Hathor International Space Brothel State Corporation of Herman, MN
- Hathor International Space Brothel State Corporation of Heron Lake, MN
- Hathor International Space Brothel State Corporation of Hewitt, MN
- Hathor International Space Brothel State Corporation of Hibbing, MN

- Hathor International Space Brothel State Corporation of Hill City, MN
- Hathor International Space Brothel State Corporation of Hills, MN
- Hathor International Space Brothel State Corporation of Hinckley, MN
- Hathor International Space Brothel State Corporation of Hitterdal, MN
- Hathor International Space Brothel State Corporation of Hoffman, MN
- Hathor International Space Brothel State Corporation of Hokah, MN
- Hathor International Space Brothel State Corporation of Holdingford, MN
- Hathor International Space Brothel State Corporation of Holland, MN
- Hathor International Space Brothel State Corporation of Hollandale, MN
- Hathor International Space Brothel State Corporation of Hopkins, MN
- Hathor International Space Brothel State Corporation of Houston, MN
- Hathor International Space Brothel State Corporation of Howard Lake, MN
- Hathor International Space Brothel State Corporation of Hoyt Lakes, MN
- Hathor International Space Brothel State Corporation of Hugo, MN
- Hathor International Space Brothel State Corporation of Hutchinson, MN
- Hathor International Space Brothel State Corporation of International Falls, MN
- Hathor International Space Brothel State Corporation of Inver Grove Heights, MN
- Hathor International Space Brothel State Corporation of Ironton, MN
- Hathor International Space Brothel State Corporation of Isanti, MN
- Hathor International Space Brothel State Corporation of Isle, MN
- Hathor International Space Brothel State Corporation of Ivanhoe, MN
- Hathor International Space Brothel State Corporation of Jackson, MN
- Hathor International Space Brothel State Corporation of Janesville, MN
- Hathor International Space Brothel State Corporation of Jasper, MN
- Hathor International Space Brothel State Corporation of Jeffers, MN
- Hathor International Space Brothel State Corporation of Jenkins, MN
- Hathor International Space Brothel State Corporation of Jordan, MN
- Hathor International Space Brothel State Corporation of Kandiyohi, MN
- Hathor International Space Brothel State Corporation of Karlstad, MN
- Hathor International Space Brothel State Corporation of Kasota, MN
- Hathor International Space Brothel State Corporation of Kasson, MN
- Hathor International Space Brothel State Corporation of Keewatin, MN
- Hathor International Space Brothel State Corporation of Kelliher, MN

- Hathor International Space Brothel State Corporation of Kellogg, MN
- Hathor International Space Brothel State Corporation of Kennedy, MN
- Hathor International Space Brothel State Corporation of Kensington, MN
- Hathor International Space Brothel State Corporation of Kenyon, MN
- Hathor International Space Brothel State Corporation of Kerkhoven, MN
- Hathor International Space Brothel State Corporation of Kiester, MN
- Hathor International Space Brothel State Corporation of Kimball, MN
- Hathor International Space Brothel State Corporation of La Crescent, MN
- Hathor International Space Brothel State Corporation of Lafayette, MN
- Hathor International Space Brothel State Corporation of Lake Benton, MN
- Hathor International Space Brothel State Corporation of Lake Bronson, MN
- Hathor International Space Brothel State Corporation of Lake City, MN
- Hathor International Space Brothel State Corporation of Lake Crystal, MN
- Hathor International Space Brothel State Corporation of Lake Elmo, MN
- Hathor International Space Brothel State Corporation of Lake Lillian, MN
- Hathor International Space Brothel State Corporation of Lake Park, MN
- Hathor International Space Brothel State Corporation of Lake Wilson, MN
- Hathor International Space Brothel State Corporation of Lakefield, MN
- Hathor International Space Brothel State Corporation of Lakeland, MN
- Hathor International Space Brothel State Corporation of Lakeville, MN
- Hathor International Space Brothel State Corporation of Lamberton, MN
- Hathor International Space Brothel State Corporation of Lancaster, MN
- Hathor International Space Brothel State Corporation of Lanesboro, MN
- Hathor International Space Brothel State Corporation of Le Center, MN
- Hathor International Space Brothel State Corporation of Le Roy, MN
- Hathor International Space Brothel State Corporation of Le Sueur, MN
- Hathor International Space Brothel State Corporation of Leota, MN
- Hathor International Space Brothel State Corporation of Lester Prairie, MN
- Hathor International Space Brothel State Corporation of Lewiston, MN
- Hathor International Space Brothel State Corporation of Lewisville, MN
- Hathor International Space Brothel State Corporation of Lindstrom, MN
- Hathor International Space Brothel State Corporation of Lismore, MN
- Hathor International Space Brothel State Corporation of Litchfield, MN

- Hathor International Space Brothel State Corporation of Little Falls, MN
- Hathor International Space Brothel State Corporation of Littlefork, MN
- Hathor International Space Brothel State Corporation of Long Lake, MN
- Hathor International Space Brothel State Corporation of Long Prairie, MN
- Hathor International Space Brothel State Corporation of Lonsdale, MN
- Hathor International Space Brothel State Corporation of Loretto, MN
- Hathor International Space Brothel State Corporation of Lowry, MN
- Hathor International Space Brothel State Corporation of Lucan, MN
- Hathor International Space Brothel State Corporation of Luverne, MN
- Hathor International Space Brothel State Corporation of Lyle, MN
- Hathor International Space Brothel State Corporation of Lynd, MN
- Hathor International Space Brothel State Corporation of Mabel, MN
- Hathor International Space Brothel State Corporation of Madelia, MN
- Hathor International Space Brothel State Corporation of Madison Lake, MN
- Hathor International Space Brothel State Corporation of Madison, MN
- Hathor International Space Brothel State Corporation of Magnolia, MN
- Hathor International Space Brothel State Corporation of Mahnomen, MN
- Hathor International Space Brothel State Corporation of Mankato, MN
- Hathor International Space Brothel State Corporation of Mantorville, MN
- Hathor International Space Brothel State Corporation of Maple Lake, MN
- Hathor International Space Brothel State Corporation of Maple Plain, MN
- Hathor International Space Brothel State Corporation of Mapleton, MN
- Hathor International Space Brothel State Corporation of Marble, MN
- Hathor International Space Brothel State Corporation of Marine On Saint Croix, MN
- Hathor International Space Brothel State Corporation of Marshall, MN
- Hathor International Space Brothel State Corporation of Mayer, MN
- Hathor International Space Brothel State Corporation of Maynard, MN
- Hathor International Space Brothel State Corporation of Mazeppa, MN
- Hathor International Space Brothel State Corporation of Mcgregor, MN
- Hathor International Space Brothel State Corporation of Mcintosh, MN
- Hathor International Space Brothel State Corporation of Medford, MN
- Hathor International Space Brothel State Corporation of Melrose, MN
- Hathor International Space Brothel State Corporation of Menahga, MN

- Hathor International Space Brothel State Corporation of Mendota, MN
- Hathor International Space Brothel State Corporation of Middle River, MN
- Hathor International Space Brothel State Corporation of Milaca, MN
- Hathor International Space Brothel State Corporation of Milan, MN
- Hathor International Space Brothel State Corporation of Milroy, MN
- Hathor International Space Brothel State Corporation of Miltona, MN
- Hathor International Space Brothel State Corporation of Minneapolis, MN
- Hathor International Space Brothel State Corporation of Minneota, MN
- Hathor International Space Brothel State Corporation of Minnesota City, MN
- Hathor International Space Brothel State Corporation of Minnesota Lake, MN
- Hathor International Space Brothel State Corporation of Minnetonka Beach, MN
- Hathor International Space Brothel State Corporation of Minnetonka, MN
- Hathor International Space Brothel State Corporation of Montevideo, MN
- Hathor International Space Brothel State Corporation of Montgomery, MN
- Hathor International Space Brothel State Corporation of Monticello, MN
- Hathor International Space Brothel State Corporation of Montrose, MN
- Hathor International Space Brothel State Corporation of Moorhead, MN
- Hathor International Space Brothel State Corporation of Moose Lake, MN
- Hathor International Space Brothel State Corporation of Mora, MN
- Hathor International Space Brothel State Corporation of Morgan, MN
- Hathor International Space Brothel State Corporation of Morris, MN
- Hathor International Space Brothel State Corporation of Morristown, MN
- Hathor International Space Brothel State Corporation of Morton, MN
- Hathor International Space Brothel State Corporation of Motley, MN
- Hathor International Space Brothel State Corporation of Mound, MN
- Hathor International Space Brothel State Corporation of Mountain Iron, MN
- Hathor International Space Brothel State Corporation of Mountain Lake, MN
- Hathor International Space Brothel State Corporation of Murdock, MN
- Hathor International Space Brothel State Corporation of Nashwauk, MN
- Hathor International Space Brothel State Corporation of Naytahwaush, MN
- Hathor International Space Brothel State Corporation of Nerstrand, MN
- Hathor International Space Brothel State Corporation of Nevis, MN
- Hathor International Space Brothel State Corporation of New Auburn, MN

- Hathor International Space Brothel State Corporation of New Germany, MN
- Hathor International Space Brothel State Corporation of New London, MN
- Hathor International Space Brothel State Corporation of New Market, MN
- Hathor International Space Brothel State Corporation of New Munich, MN
- Hathor International Space Brothel State Corporation of New Prague, MN
- Hathor International Space Brothel State Corporation of New Richland, MN
- Hathor International Space Brothel State Corporation of New Ulm, MN
- Hathor International Space Brothel State Corporation of New York Mills, MN
- Hathor International Space Brothel State Corporation of Newfolden, MN
- Hathor International Space Brothel State Corporation of Newport, MN
- Hathor International Space Brothel State Corporation of Nicollet, MN
- Hathor International Space Brothel State Corporation of Nisswa, MN
- Hathor International Space Brothel State Corporation of North Branch, MN
- Hathor International Space Brothel State Corporation of Northfield, MN
- Hathor International Space Brothel State Corporation of Northome, MN
- Hathor International Space Brothel State Corporation of Northrop, MN
- Hathor International Space Brothel State Corporation of Norwood, MN
- Hathor International Space Brothel State Corporation of Oak Park, MN
- Hathor International Space Brothel State Corporation of Ogilvie, MN
- Hathor International Space Brothel State Corporation of Oklee, MN
- Hathor International Space Brothel State Corporation of Olivia, MN
- Hathor International Space Brothel State Corporation of Onamia, MN
- Hathor International Space Brothel State Corporation of Oronoco, MN
- Hathor International Space Brothel State Corporation of Orr, MN
- Hathor International Space Brothel State Corporation of Ortonville, MN
- Hathor International Space Brothel State Corporation of Osakis, MN
- Hathor International Space Brothel State Corporation of Oslo, MN
- Hathor International Space Brothel State Corporation of Osseo, MN
- Hathor International Space Brothel State Corporation of Ostrander, MN
- Hathor International Space Brothel State Corporation of Ottertail, MN
- Hathor International Space Brothel State Corporation of Owatonna, MN
- Hathor International Space Brothel State Corporation of Park Rapids, MN
- Hathor International Space Brothel State Corporation of Parkers Prairie, MN

- Hathor International Space Brothel State Corporation of Paynesville, MN
- Hathor International Space Brothel State Corporation of Pelican Rapids, MN
- Hathor International Space Brothel State Corporation of Pemberton, MN
- Hathor International Space Brothel State Corporation of Pennock, MN
- Hathor International Space Brothel State Corporation of Pequot Lakes, MN
- Hathor International Space Brothel State Corporation of Perham, MN
- Hathor International Space Brothel State Corporation of Peterson, MN
- Hathor International Space Brothel State Corporation of Pierz, MN
- Hathor International Space Brothel State Corporation of Pillager, MN
- Hathor International Space Brothel State Corporation of Pine City, MN
- Hathor International Space Brothel State Corporation of Pine Island, MN
- Hathor International Space Brothel State Corporation of Pine River, MN
- Hathor International Space Brothel State Corporation of Pipestone, MN
- Hathor International Space Brothel State Corporation of Plainview, MN
- Hathor International Space Brothel State Corporation of Plato, MN
- Hathor International Space Brothel State Corporation of Plummer, MN
- Hathor International Space Brothel State Corporation of Ponemah, MN
- Hathor International Space Brothel State Corporation of Preston, MN
- Hathor International Space Brothel State Corporation of Princeton, MN
- Hathor International Space Brothel State Corporation of Prinsburg, MN
- Hathor International Space Brothel State Corporation of Prior Lake, MN
- Hathor International Space Brothel State Corporation of Racine, MN
- Hathor International Space Brothel State Corporation of Randall, MN
- Hathor International Space Brothel State Corporation of Randolph, MN
- Hathor International Space Brothel State Corporation of Raymond, MN
- Hathor International Space Brothel State Corporation of Red Lake Falls, MN
- Hathor International Space Brothel State Corporation of Red Wing, MN
- Hathor International Space Brothel State Corporation of Redby, MN
- Hathor International Space Brothel State Corporation of Redlake, MN
- Hathor International Space Brothel State Corporation of Redwood Falls, MN
- Hathor International Space Brothel State Corporation of Remer, MN
- Hathor International Space Brothel State Corporation of Renville, MN
- Hathor International Space Brothel State Corporation of Rice, MN

- Hathor International Space Brothel State Corporation of Richmond, MN
- Hathor International Space Brothel State Corporation of Rochester, MN
- Hathor International Space Brothel State Corporation of Rock Creek, MN
- Hathor International Space Brothel State Corporation of Rockford, MN
- Hathor International Space Brothel State Corporation of Rockville, MN
- Hathor International Space Brothel State Corporation of Rogers, MN
- Hathor International Space Brothel State Corporation of Rollingstone, MN
- Hathor International Space Brothel State Corporation of Rose Creek, MN
- Hathor International Space Brothel State Corporation of Roseau, MN
- Hathor International Space Brothel State Corporation of Rosemount, MN
- Hathor International Space Brothel State Corporation of Rothsay, MN
- Hathor International Space Brothel State Corporation of Round Lake, MN
- Hathor International Space Brothel State Corporation of Royalton, MN
- Hathor International Space Brothel State Corporation of Rush City, MN
- Hathor International Space Brothel State Corporation of Rushford, MN
- Hathor International Space Brothel State Corporation of Rushmore, MN
- Hathor International Space Brothel State Corporation of Russell, MN
- Hathor International Space Brothel State Corporation of Ruthton, MN
- Hathor International Space Brothel State Corporation of Sabin, MN
- Hathor International Space Brothel State Corporation of Sacred Heart, MN
- Hathor International Space Brothel State Corporation of Saint Bonifacius, MN
- Hathor International Space Brothel State Corporation of Saint Charles, MN
- Hathor International Space Brothel State Corporation of Saint Clair
- Hathor International Space Brothel State Corporation of Saint Cloud, MN
- Hathor International Space Brothel State Corporation of Saint Francis, MN
- Hathor International Space Brothel State Corporation of Saint Hilaire, MN
- Hathor International Space Brothel State Corporation of Saint James, MN
- Hathor International Space Brothel State Corporation of Saint Joseph, MN
- Hathor International Space Brothel State Corporation of Saint Martin, MN
- Hathor International Space Brothel State Corporation of Saint Michael, MN
- Hathor International Space Brothel State Corporation of Saint Paul Park, MN
- Hathor International Space Brothel State Corporation of Saint Paul, MN
- Hathor International Space Brothel State Corporation of Saint Peter, MN

- Hathor International Space Brothel State Corporation of Saint Stephen, MN
- Hathor International Space Brothel State Corporation of Sanborn, MN
- Hathor International Space Brothel State Corporation of Sandstone, MN
- Hathor International Space Brothel State Corporation of Sartell, MN
- Hathor International Space Brothel State Corporation of Sauk Centre, MN
- Hathor International Space Brothel State Corporation of Sauk Rapids, MN
- Hathor International Space Brothel State Corporation of Savage, MN
- Hathor International Space Brothel State Corporation of Sebeka, MN
- Hathor International Space Brothel State Corporation of Shafer, MN
- Hathor International Space Brothel State Corporation of Shakopee, MN
- Hathor International Space Brothel State Corporation of Shelly, MN
- Hathor International Space Brothel State Corporation of Sherburn, MN
- Hathor International Space Brothel State Corporation of Silver Bay, MN
- Hathor International Space Brothel State Corporation of Silver Lake, MN
- Hathor International Space Brothel State Corporation of Slayton, MN
- Hathor International Space Brothel State Corporation of Sleepy Eye, MN
- Hathor International Space Brothel State Corporation of South Haven, MN
- Hathor International Space Brothel State Corporation of South Saint Paul, MN
- Hathor International Space Brothel State Corporation of Spicer, MN
- Hathor International Space Brothel State Corporation of Spring Grove, MN
- Hathor International Space Brothel State Corporation of Spring Lake, MN
- Hathor International Space Brothel State Corporation of Spring Park, MN
- Hathor International Space Brothel State Corporation of Spring Valley, MN
- Hathor International Space Brothel State Corporation of Springfield, MN
- Hathor International Space Brothel State Corporation of Stacy, MN
- Hathor International Space Brothel State Corporation of Staples, MN
- Hathor International Space Brothel State Corporation of Starbuck, MN
- Hathor International Space Brothel State Corporation of Stephen, MN
- Hathor International Space Brothel State Corporation of Stewart, MN
- Hathor International Space Brothel State Corporation of Stewartville, MN
- Hathor International Space Brothel State Corporation of Stillwater, MN
- Hathor International Space Brothel State Corporation of Stockton, MN
- Hathor International Space Brothel State Corporation of Storden, MN

- Hathor International Space Brothel State Corporation of Sturgeon Lake, MN
- Hathor International Space Brothel State Corporation of Swanville, MN
- Hathor International Space Brothel State Corporation of Taconite, MN
- Hathor International Space Brothel State Corporation of Taunton, MN
- Hathor International Space Brothel State Corporation of Taylors Falls, MN
- Hathor International Space Brothel State Corporation of Thief River Falls, MN
- Hathor International Space Brothel State Corporation of Tower, MN
- Hathor International Space Brothel State Corporation of Tracy, MN
- Hathor International Space Brothel State Corporation of Trimont, MN
- Hathor International Space Brothel State Corporation of Truman, MN
- Hathor International Space Brothel State Corporation of Twin Valley, MN
- Hathor International Space Brothel State Corporation of Two Harbors, MN
- Hathor International Space Brothel State Corporation of Tyler, MN
- Hathor International Space Brothel State Corporation of Ulen, MN
- Hathor International Space Brothel State Corporation of Underwood, MN
- Hathor International Space Brothel State Corporation of Upsala, MN
- Hathor International Space Brothel State Corporation of Utica, MN
- Hathor International Space Brothel State Corporation of Vergas, MN
- Hathor International Space Brothel State Corporation of Vermillion, MN
- Hathor International Space Brothel State Corporation of Verndale, MN
- Hathor International Space Brothel State Corporation of Vernon Center, MN
- Hathor International Space Brothel State Corporation of Vesta, MN
- Hathor International Space Brothel State Corporation of Victoria, MN
- Hathor International Space Brothel State Corporation of Villard, MN
- Hathor International Space Brothel State Corporation of Virginia, MN
- Hathor International Space Brothel State Corporation of Wabasha, MN
- Hathor International Space Brothel State Corporation of Wabasso, MN
- Hathor International Space Brothel State Corporation of Waconia, MN
- Hathor International Space Brothel State Corporation of Wadena, MN
- Hathor International Space Brothel State Corporation of Wahkon, MN
- Hathor International Space Brothel State Corporation of Waite Park, MN
- Hathor International Space Brothel State Corporation of Waldorf, MN
- Hathor International Space Brothel State Corporation of Walker, MN

- Hathor International Space Brothel State Corporation of Walnut Grove, MN
- Hathor International Space Brothel State Corporation of Wanamingo, MN
- Hathor International Space Brothel State Corporation of Warren, MN
- Hathor International Space Brothel State Corporation of Warroad, MN
- Hathor International Space Brothel State Corporation of Waseca, MN
- Hathor International Space Brothel State Corporation of Watertown, MN
- Hathor International Space Brothel State Corporation of Waterville, MN
- Hathor International Space Brothel State Corporation of Watkins, MN
- Hathor International Space Brothel State Corporation of Watson, MN
- Hathor International Space Brothel State Corporation of Waubun, MN
- Hathor International Space Brothel State Corporation of Waverly, MN
- Hathor International Space Brothel State Corporation of Wayzata, MN
- Hathor International Space Brothel State Corporation of Welcome, MN
- Hathor International Space Brothel State Corporation of Wells, MN
- Hathor International Space Brothel State Corporation of West Concord, MN
- Hathor International Space Brothel State Corporation of Westbrook, MN
- Hathor International Space Brothel State Corporation of Wheaton, MN
- Hathor International Space Brothel State Corporation of White Earth, MN
- Hathor International Space Brothel State Corporation of Willernie, MN
- Hathor International Space Brothel State Corporation of Williams, MN
- Hathor International Space Brothel State Corporation of Willmar, MN
- Hathor International Space Brothel State Corporation of Willow River, MN
- Hathor International Space Brothel State Corporation of Wilmont, MN
- Hathor International Space Brothel State Corporation of Windom, MN
- Hathor International Space Brothel State Corporation of Winger, MN
- Hathor International Space Brothel State Corporation of Winnebago, MN
- Hathor International Space Brothel State Corporation of Winona, MN
- Hathor International Space Brothel State Corporation of Winsted, MN
- Hathor International Space Brothel State Corporation of Winthrop, MN
- Hathor International Space Brothel State Corporation of Wood Lake, MN
- Hathor International Space Brothel State Corporation of Worthington, MN
- Hathor International Space Brothel State Corporation of Wrenshall, MN
- Hathor International Space Brothel State Corporation of Wykoff, MN

- **Hathor International Space Brothel State Corporation of Wyoming, MN**
- **Hathor International Space Brothel State Corporation of Zimmerman, MN**
- **Hathor International Space Brothel State Corporation of Zumbrota, MN**

There are a total of 912 cities in Minnesota. By selling containers to each city through this corporation, the projected profit sales amount to $54,720,000,000 annually. To fulfill the Minnesota market, we need 52 International Space State Corporations.

- People might think it's unconventional, but...
- We are the people who live in the real world and see it happening, even if it's illegal.
- We are the people who believe that directing this money to the proper use can bring free healthcare, education, and other benefits to the people.
- We are the people who believe that, through this, we can contribute to space research and colonization of planets.
- We are the people who believe in the prosperity of countries and the world.

To develop and present this project to the public, we need:

1. Financing
2. Selection of professional employees, specialists
3. Selection and purchase of special materials, technologies for development
4. Open government facilities for public access

Revenue from the sales will be directed towards healthcare, military agency expenses, and the space project program created by Georgiy S Garbuz. This program aims to build a national Garbuz Space Academy and national Garbuz Space Academy schools worldwide to grant foster care and other kids access to education. Funding for grants to educate these kids will be derived from directed taxes on Georgiy S Garbuz's science works and all projects.

With the provision of sex services through this partner corporation, the total 912 cities in Minnesota yield a profit of $2,190,987,000,000 annually. To fulfill the Minnesota market, we estimate the need for 4 International Space State corporations with 3,648 partner brothels, employing over a million sex workers.

The project was initially conceived by Georgiy Sergeyevich Garbuz in 2007 and is being developed under the name "Project For The Future" (G.S. Garbuz) at the National Garbuz Space Academy Trust Corporation

---

PROJECT HEMKEN PENANCE

Project Hemken Penance is an international government program designed to facilitate the transition of illegal prostitution into a legal and regulated industry worldwide. The program aims to legalize former pimps and prostitutes, transforming them into legal business entities after meeting all government agency requirements. With the support of the president, our goal is to legalize sex services in the United States, making it a legitimate market.

Currently, individuals engaged in illegal sex trade operate in the black market and face legal consequences. The Hemken Penance program provides an avenue for these individuals to become legal business operators, complying with government regulations and contributing positively to the country's economy. The program collaborates with law enforcement agencies such as the police, FBI, CIA, Interpol, and others globally.

Former pimps and prostitutes participating in the Hemken Penance program are required to confess to engaging in illegal sex services and other related crimes. This information, including details of blackmail, bribes, kidnapping, and other criminal activities, is used by government agencies to analyze, work, and potentially recruit individuals around the world.

The Hemken Penance program enables former sex workers to legalize their businesses, fostering a legal and profitable environment within the country. The program's success relies on public support and the belief that redirecting funds from this industry can contribute to free healthcare, education, space exploration, and overall prosperity.

To implement the Hemken Penance program publicly, we are:

1. Realists who have witnessed the illegal sex trade's impact on society.
2. Advocates for redirecting funds for societal betterment.
3. Believers in the transformative power of legalizing and regulating the sex trade.

For the successful development and execution of the Hemken Penance program, we require:

1. Financing
2. Selection of professional employees and specialists
3. Selection and purchase of special materials and technologies for development
4. Opening government facilities for public transparency

Revenue generated from the Hemken Penance program's sales will be allocated to healthcare, military agency expenses, and the space project program initiated by Georgiy S. Garbuz. This program aims to establish the National Garbuz Space Academy and affiliated schools globally, providing grants to foster care and underprivileged children for their education. Funding for these grants will be sourced from directed taxes on Georgiy S. Garbuz's science works and all associated projects.

The project was initially conceived by Georgiy Sergeyevich Garbuz in 2007 and is being developed under the name "Project For The Future" (G.S. Garbuz) at the National Garbuz Space Academy Trust Corporation

---

CONTAINER TO THE CITY

The Container to the City initiative is a government-partner drug program designed to legalize and wholesale non-addictive marijuana in every country. This innovative program aims to boost the economy of cities and countries alike. Drawing inspiration from the successful legalization of marijuana in the United States since 2007, our research indicates the potential for economic growth by legalizing marijuana in regions with struggling economies. This initiative also includes a unique space-state corporation model to efficiently manage the wholesale distribution of marijuana.

To fulfill the market in Minnesota, we estimate a need for 46,436 containers, each containing around 40,000 pounds of marijuana. The sales have consistently increased since the legalization, making this an excellent opportunity to purchase from countries with weaker economies. The international program is designed to extend economic support to regions such as Esik (Kazakhstan), Novaia Sinjereia (Moldova), and Chernovchi (Ukraine), where prices per kilogram on the black market were historically low. By selling at a higher price of $2,989.99 per kilogram in Minnesota, we anticipate significant profits, contributing to the overall economic growth.

A unique aspect of this program involves the establishment of a government-partner space-state corporation, owned by government agencies and citizens-turned-businessmen. Each container requires a permit, generating substantial profits and bolstering the economy. We propose allocating one container per permission to city and state corporations across the USA.

This strategic move supports cities' economic development and facilitates the transition of former drug dealers into legal businessmen, creating partner wholesale stores in cities and countries.

The Container to the City program envisions redirecting profits towards essential public services, including healthcare, education, and other crucial needs. The initiative also aligns with our broader vision to fund space projects, research, and colonization efforts led by Georgiy S. Garbuz.

Implementation and Development: To ensure the success of this groundbreaking program, the following steps are crucial:

1. **Financing:** Secure adequate financial support for program initiation and continued development.
2. **Selection of Professional Employees and Specialists:** Assemble a team of experts to oversee the program's efficient functioning and growth.
3. **Selection and Purchase of Special Materials and Technologies:** Acquire cutting-edge materials and technologies to streamline container operations.
4. **Open Government Facilities for Public Access:** Establish accessible government facilities to foster public engagement and support.

Revenue generated from the program will be directed towards essential public services, space projects, and education. This includes funding grants to educate foster care children and support various projects initiated by Georgiy S. Garbuz.

Container to the City is a visionary initiative that leverages the economic potential of legalized marijuana to benefit cities, countries, and individuals. By embracing this program, we aim to create a sustainable model that not only stimulates economic growth but also channels funds toward vital societal needs and advancements in space exploration.

The project was initially conceived by Georgiy Sergeyevich Garbuz in 2007 and is being developed under the name "Project For The Future" (G.S. Garbuz) at the National Garbuz Space Academy Trust Corporation

---

INTERNATIONAL GOVERNMENT DRUG CONFESSION PROGRAM

The International Government Drug Confession Program marks a significant step towards transitioning illegal, non-addictive drug marijuana into a legal market worldwide. This initiative aims to legalize former drug dealers, transforming them into legal businessmen by meeting all government requirements. Non-addictive marijuana has been legalized in the United States, opening up opportunities for legal sales. This transition has implications globally, especially for those operating in the black market, now moving towards legitimacy.

The drug confession program facilitates the legalization of non-addictive marijuana, allowing former drug dealers to become legal businessmen. The program emphasizes cooperation with law enforcement agencies such as the police, FBI, CIA, Interpol, and others across the globe. By compelling former drug dealers to confess to their illegal activities, including blackmail, bribes, kidnapping, and other crimes associated with illegal marijuana trade, the government agencies gain valuable information to analyze and recruit individuals worldwide.

Former drug dealers can now transition into legal business ventures, contributing to the country's economy and prospering within the legal framework. The drug confession program not only offers economic opportunities but also addresses public safety concerns associated with illegal drug activities. The collaboration with law enforcement ensures a systematic approach to managing the transition and bringing former offenders into the legal fold.

To introduce the drug confession program to the public, it is essential to garner support from citizens who believe in redirecting funds for constructive purposes. Acknowledging the skepticism, the program emphasizes its potential to fund critical areas such as healthcare, education, and space exploration. Implementing the program requires financing, a team of professionals, special materials, technologies, and government facilities open to the public.

The drug confession program envisions a future where funds generated from legal drug sales contribute to societal welfare, education, and advancements in space exploration. The program aligns with the broader vision of Georgiy S. Garbuz, directing revenues towards space projects and the creation of national space academies and schools.

The International Drug Confession Program signifies a bold move towards marijuana legalization, offering a chance for former drug dealers to reintegrate into society as legal businessmen. By embracing this initiative, we not only address the economic potential of legal drug sales but also pave the way for societal development and exploration of new frontiers in space. The program's success hinges on public support, effective financing, and strategic collaboration with law enforcement agencies.

The project was initially conceived by Georgiy Sergeyevich Garbuz in 2007 and is being developed under the name "Project For The Future" (G.S. Garbuz) at the National Garbuz Space Academy Trust Corporation

---

DEMETRA SPACE FARM GOVERNMENT PROGRAM

The Demetra Space Government Program is a groundbreaking initiative aimed at boosting agricultural activities in impoverished countries and around the world. By facilitating economic growth and improving family budgets, this program focuses on increasing population engagement in government-owned farms. The strategic partnership involves international intergalactic space farm corporations jointly owned by the government and the people.

1. **Ownership Requirement:**

- Farmers are mandated to own a minimum of 50% of the farm, ensuring responsible management.
- Educated individuals with a background in agriculture are encouraged to serve as general managers, overseeing day-to-day operations.

2. **Farm Structure:**

- Each farm is an individual entity under the government corporation, fostering self-sufficiency and prosperity for citizens worldwide.
- Farms are designed to generate one million dollars annually through the cultivation of fruits and vegetables on 100 acres of land.

3. Marijuana Project:

- To enhance economic opportunities, the program incorporates the cultivation of marijuana in impoverished countries such as Ukraine, Moldova, Kazakhstan, Russia, and others.
- For marijuana projects, farms must allocate 10 acres, with 50% ownership by the government and 50% by local citizen farmers, allowing for annual revenue generation and export to countries like America and Europe.

To introduce the Demetra Space Government Program to the public, it is crucial to garner support from individuals who understand the potential economic, social, and health benefits. Despite potential skepticism, the program emphasizes redirecting funds for positive purposes, such as healthcare, education, and space exploration. Publicly accessible government facilities will be crucial for transparency.

The Demetra Space Government Program envisions a future where the generated revenue significantly contributes to societal welfare, healthcare, military agencies, and advancements in space exploration. By channeling funds from agricultural initiatives, the program aligns with the broader vision of Georgiy S. Garbuz, focusing on education and fostering the well-being of foster care and other children globally.

The Demetra Space Government Program represents a visionary approach to global agricultural development, emphasizing the integration of impoverished countries into the economic mainstream. Success in implementing this program relies on securing public support, effective financing, and strategic partnerships. Through responsible and sustainable agricultural practices, this initiative aims to uplift communities, strengthen economies, and propel humanity towards a brighter future.

The project was initially conceived by Georgiy Sergeyevich Garbuz in 2007 and is being developed under the name "Project For The Future" (G.S. Garbuz) at the National Garbuz Space Academy Trust Corporation

---

WALZ INTERNATIONAL SPACE DRUG STATE CORPORATION OF MINNESOTA

The International Intergalactic State Corporation is a visionary initiative aimed at bolstering the economy of nations, states, and the world while delving into the exploration of space. With the support of the president, we successfully played a pivotal role in legalizing non-addictive drugs in the United States, and we aspire to extend this impact by advocating for the global legalization of marijuana.

Our program is strategically designed to purchase and cultivate marijuana in countries facing severe economic challenges. Reflecting on 2003 prices in cities like Esik (Kazakhstan), Novaia Sinjereia (Moldova), and Chernovchi (Ukraine), where marijuana was priced between $5-$10 per kilogram, we aim to harness the economic potential of this plant. In contrast, the same quantity was valued at $3000 in the United States, illustrating the significant profitability in the black market.

Based on our research, one container to the city is projected to be sold within a week, generating an annual profit of $2.84 trillion. To meet the needs of the state of Minnesota, 52 International State Corporations are required, contributing a staggering $147.96 trillion in annual profit. The transition from illegal to legal sales presents an opportunity for state corporations to control the drug market, allowing individuals to open partner stores in every city.

Scaling our operations to meet the demand of the entire USA market entails the production of 2.37 million containers, each weighing 40,000 pounds. With each state having control over its drug market through state corporations, individuals can actively participate by opening partner stores in cities.

Recognizing the shift from illegal to legal sales, a special program has been initiated to facilitate the integration of former drug dealers into legal businesses. This program requires public acceptance, enabling individuals to work legally and participate in the legal sale of marijuana across the country.

The International Intergalactic State Corporation is at the forefront of transformative economic and space exploration initiatives. By leveraging the economic potential of marijuana cultivation and embracing the legalization transition, our program envisions a future where nations prosper economically and humanity expands its horizons in the exploration of space.

Walz International Space Drug State Corporation of Minnesota

- 40%: State Government
- 20%: Imperial Space Federation (USA Government)
- 10%: Federal Government
- 10%: NASA
- 5%: CIA, Interpol, FBI, or Military
- 5%: State Police or Federal Police
- 5%: Russia, Moldova, Mexico, or any other country government
- 5%: Igor, Local Businessmen, or Former Drug Dealer (subject to meeting all government agency requirements)

Walz International Space Drug State Corporation of Ada, MN

- 40%: Walz International Space Drug State Corporation of Minnesota (Owner)

- 10%: Imperial Space Federation
- 50%: Local Businessmen or Former Drug Dealer (subject to meeting all government agency requirements)

Walz International Space Drug State Corporation of Adams, MN

- 40%: Walz International Space Drug State Corporation of Minnesota (Owner)
- 10%: Imperial Space Federation
- 50%: Local Businessmen or Former Drug Dealer (subject to meeting all government agency requirements)

Walz International Space Drug State Corporation of Adrian, MN

- 40%: Walz International Space Drug State Corporation of Minnesota (Owner)
- 10%: Imperial Space Federation
- 50%: Local Businessmen or Former Drug Dealer (subject to meeting all government agency requirements)

Walz International Space Drug State Corporation of Afton, MN

- 40%: Walz International Space Drug State Corporation of Minnesota (Owner)
- 10%: Imperial Space Federation
- 50%: Local Businessmen or Former Drug Dealer (subject to meeting all government agency requirements)

Walz International Space Drug State Corporation of Aitkin, MN

- 40%: Walz International Space Drug State Corporation of Minnesota (Owner)
- 10%: Imperial Space Federation

- 50%: Local Businessmen or Former Drug Dealer (subject to meeting all government agency requirements)

Walz International Space Drug State Corporation of Akeley, MN

- 40%: Walz International Space Drug State Corporation of Minnesota (Owner)
- 10%: Imperial Space Federation
- 50%: Local Businessmen or Former Drug Dealer (subject to meeting all government agency requirements)

Walz International Space Drug State Corporation of Albany, MN

- 40%: Walz International Space Drug State Corporation of Minnesota (Owner)
- 10%: Imperial Space Federation
- 50%: Local Businessmen or Former Drug Dealer (subject to meeting all government agency requirements)

Walz International Space Drug State Corporation of Albert Lea, MN

- 40%: Walz International Space Drug State Corporation of Minnesota (Owner)
- 10%: Imperial Space Federation
- 50%: Local Businessmen or Former Drug Dealer (subject to meeting all government agency requirements)

Walz International Space Drug State Corporation of Albertville, MN

- 40%: Walz International Space Drug State Corporation of Minnesota (Owner)
- 10%: Imperial Space Federation

- 50%: Local Businessmen or Former Drug Dealer (subject to meeting all government agency requirements)

Walz International Space Drug State Corporation of Alden, MN

- 40%: Walz International Space Drug State Corporation of Minnesota (Owner)
- 10%: Imperial Space Federation
- 50%: Local Businessmen or Former Drug Dealer (subject to meeting all government agency requirements)

Walz International Space Drug State Corporation of Alexandria, MN

- 40%: Walz International Space Drug State Corporation of Minnesota (Owner)
- 10%: Imperial Space Federation
- 50%: Local Businessmen or Former Drug Dealer (subject to meeting all government agency requirements)

Walz International Space Drug State Corporation of Aitura, MN

- 40%: Walz International Space Drug State Corporation of Minnesota (Owner)
- 10%: Imperial Space Federation
- 50%: Local Businessmen or Former Drug Dealer (subject to meeting all government agency requirements)

Walz International Space Drug State Corporation of Alvarado, MN

- 40%: Walz International Space Drug State Corporation of Minnesota (Owner)
- 10%: Imperial Space Federation

- 50%: Local Businessmen or Former Drug Dealer (subject to meeting all government agency requirements)

Walz International Space Drug State Corporation of Amboy, MN

- 40%: Walz International Space Drug State Corporation of Minnesota (Owner)
- 10%: Imperial Space Federation
- 50%: Local Businessmen or Former Drug Dealer (subject to meeting all government agency requirements)

Walz International Space Drug State Corporation of Andover, MN

- 40%: Walz International Space Drug State Corporation of Minnesota (Owner)
- 10%: Imperial Space Federation
- 50%: Local Businessmen or Former Drug Dealer (subject to meeting all government agency requirements)

Walz International Space Drug State Corporation of Annandale, MN

- 40%: Walz International Space Drug State Corporation of Minnesota (Owner)
- 10%: Imperial Space Federation
- 50%: Local Businessmen or Former Drug Dealer (subject to meeting all government agency requirements)

Walz International Space Drug State Corporation of Anoka, MN

- 40%: Walz International Space Drug State Corporation of Minnesota (Owner)
- 10%: Imperial Space Federation

- 50%: Local Businessmen or Former Drug Dealer (subject to meeting all government agency requirements)

Walz International Space Drug State Corporation of Appleton, MN

- 40%: Walz International Space Drug State Corporation of Minnesota (Owner)
- 10%: Imperial Space Federation
- 50%: Local Businessmen or Former Drug Dealer (subject to meeting all government agency requirements)

Walz International Space Drug State Corporation of Argyle, MN

- 40%: Walz International Space Drug State Corporation of Minnesota (Owner)
- 10%: Imperial Space Federation
- 50%: Local Businessmen or Former Drug Dealer (subject to meeting all government agency requirements)

Walz International Space Drug State Corporation of Arlington, MN

- 40%: Walz International Space Drug State Corporation of Minnesota (Owner)
- 10%: Imperial Space Federation
- 50%: Local Businessmen or Former Drug Dealer (subject to meeting all government agency requirements)

Walz International Space Drug State Corporation of Ashby, MN

- 40%: Walz International Space Drug State Corporation of Minnesota (Owner)
- 10%: Imperial Space Federation

- 50%: Local Businessmen or Former Drug Dealer (subject to meeting all government agency requirements)

Walz International Space Drug State Corporation of Askov, MN

- 40%: Walz International Space Drug State Corporation of Minnesota (Owner)
- 10%: Imperial Space Federation
- 50%: Local Businessmen or Former Drug Dealer (subject to meeting all government agency requirements)

Walz International Space Drug State Corporation of Atwater, MN

- 40%: Walz International Space Drug State Corporation of Minnesota (Owner)
- 10%: Imperial Space Federation
- 50%: Local Businessmen or Former Drug Dealer (subject to meeting all government agency requirements)

Walz International Space Drug State Corporation of Audubon, MN

- 40%: Walz International Space Drug State Corporation of Minnesota (Owner)
- 10%: Imperial Space Federation
- 50%: Local Businessmen or Former Drug Dealer (subject to meeting all government agency requirements)

In addition, Walz International Space Drug State Corporations will be created for the other following cities in the state of Minnesota, operating with the same ownership structure as above:

- **Walz International Space Drug State Corporation of Aurora, MN**
- **Walz International Space Drug State Corporation of Austin, MN**
- **Walz International Space Drug State Corporation of Avon, MN**

- **Walz International Space Drug State Corporation of Babbitt, MN**
- **Walz International Space Drug State Corporation of Backus, MN**
- **Walz International Space Drug State Corporation of Badger, MN**
- **Walz International Space Drug State Corporation of Bagley, MN**
- **Walz International Space Drug State Corporation of Balaton, MN**
- **Walz International Space Drug State Corporation of Barnesville, MN**
- **Walz International Space Drug State Corporation of Barnum, MN**
- **Walz International Space Drug State Corporation of Barrett, MN**
- **Walz International Space Drug State Corporation of Battle Lake, MN**
- **Walz International Space Drug State Corporation of Baudette, MN**
- **Walz International Space Drug State Corporation of Baxter, MN**
- **Walz International Space Drug State Corporation of Bayport, MN**
- **Walz International Space Drug State Corporation of Beardlsey, MN**
- **Walz International Space Drug State Corporation of Beaver Creek, MN**
- **Walz International Space Drug State Corporation of Becker, MN**
- **Walz International Space Drug State Corporation of Belgrade, MN**
- **Walz International Space Drug State Corporation of Belle Plaine, MN**
- **Walz International Space Drug State Corporation of Bellingham, MN**
- **Walz International Space Drug State Corporation of Belview, MN**
- **Walz International Space Drug State Corporation of Bemidji, MN**
- **Walz International Space Drug State Corporation of Benson, MN**
- **Walz International Space Drug State Corporation of Bertha, MN**
- **Walz International Space Drug State Corporation of Bethel, MN**
- **Walz International Space Drug State Corporation of Big Falls, MN**
- **Walz International Space Drug State Corporation of Big Lake, MN**
- **Walz International Space Drug State Corporation of Bigelow, MN**
- **Walz International Space Drug State Corporation of Bigfork, MN**
- **Walz International Space Drug State Corporation of Bird Island, MN**
- **Walz International Space Drug State Corporation of Biwabik, MN**
- **Walz International Space Drug State Corporation of Blackduck, MN**
- **Walz International Space Drug State Corporation of Blooming Prairie, MN**
- **Walz International Space Drug State Corporation of Blue Earth, MN**
- **Walz International Space Drug State Corporation of Bluffton, MN**

- Walz International Space Drug State Corporation of Bovey, MN
- Walz International Space Drug State Corporation of Bowlus, MN
- Walz International Space Drug State Corporation of Boyd, MN
- Walz International Space Drug State Corporation of Braham, MN
- Walz International Space Drug State Corporation of Brainerd, MN
- Walz International Space Drug State Corporation of Brandon, MN
- Walz International Space Drug State Corporation of Breckenridge, MN
- Walz International Space Drug State Corporation of Brewster, MN
- Walz International Space Drug State Corporation of Bricelyn, MN
- Walz International Space Drug State Corporation of Brooten, MN
- Walz International Space Drug State Corporation of Browerville, MN
- Walz International Space Drug State Corporation of Browns Valley, MN
- Walz International Space Drug State Corporation of Brownsdale, MN
- Walz International Space Drug State Corporation of Brownsville, MN
- Walz International Space Drug State Corporation of Brownton, MN
- Walz International Space Drug State Corporation of Buckman, MN
- Walz International Space Drug State Corporation of Buffalo Lake, MN
- Walz International Space Drug State Corporation of Buffalo, MN
- Walz International Space Drug State Corporation of Buhl, MN
- Walz International Space Drug State Corporation of Burnsville, MN
- Walz International Space Drug State Corporation of Butterfield, MN
- Walz International Space Drug State Corporation of Byron, MN
- Walz International Space Drug State Corporation of Caledonia, MN
- Walz International Space Drug State Corporation of Calumet, MN
- Walz International Space Drug State Corporation of Cambridge, MN
- Walz International Space Drug State Corporation of Campbell, MN
- Walz International Space Drug State Corporation of Canby, MN
- Walz International Space Drug State Corporation of Cannon Falls, MN
- Walz International Space Drug State Corporation of Canton, MN
- Walz International Space Drug State Corporation of Carlos, MN
- Walz International Space Drug State Corporation of Carlton, MN
- Walz International Space Drug State Corporation of Carver, MN
- Walz International Space Drug State Corporation of Cass Lake, MN

- Walz International Space Drug State Corporation of Center City, MN
- Walz International Space Drug State Corporation of Ceylon, MN
- Walz International Space Drug State Corporation of Champlin, MN
- Walz International Space Drug State Corporation of Chandler, MN
- Walz International Space Drug State Corporation of Chanhassen, MN
- Walz International Space Drug State Corporation of Chaska, MN
- Walz International Space Drug State Corporation of Chatfield, MN
- Walz International Space Drug State Corporation of Chisago City, MN
- Walz International Space Drug State Corporation of Chisholm, MN
- Walz International Space Drug State Corporation of Chokio, MN
- Walz International Space Drug State Corporation of Circle Pines, MN
- Walz International Space Drug State Corporation of Clara City, MN
- Walz International Space Drug State Corporation of Claremont, MN
- Walz International Space Drug State Corporation of Clarissa, MN
- Walz International Space Drug State Corporation of Clarkfield, MN
- Walz International Space Drug State Corporation of Clarks Grove, MN
- Walz International Space Drug State Corporation of Clear Lake, MN
- Walz International Space Drug State Corporation of Clearbrook, MN
- Walz International Space Drug State Corporation of Clearwater, MN
- Walz International Space Drug State Corporation of Cleveland, MN
- Walz International Space Drug State Corporation of Climax, MN
- Walz International Space Drug State Corporation of Clinton, MN
- Walz International Space Drug State Corporation of Cloquet, MN
- Walz International Space Drug State Corporation of Cohasset, MN
- Walz International Space Drug State Corporation of Cokato, MN
- Walz International Space Drug State Corporation of Cold Spring, MN
- Walz International Space Drug State Corporation of Coleraine, MN
- Walz International Space Drug State Corporation of Cologne, MN
- Walz International Space Drug State Corporation of Comfrey, MN
- Walz International Space Drug State Corporation of Cook, MN
- Walz International Space Drug State Corporation of Cosmos, MN
- Walz International Space Drug State Corporation of Cottage Grove, MN
- Walz International Space Drug State Corporation of Cottonwood, MN
- Walz International Space Drug State Corporation of Courtland, MN
- Walz International Space Drug State Corporation of Crookston, MN

- Walz International Space Drug State Corporation of Crosby, MN
- Walz International Space Drug State Corporation of Crosslake, MN
- Walz International Space Drug State Corporation of Crystal Bay, MN
- Walz International Space Drug State Corporation of Currie, MN
- Walz International Space Drug State Corporation of Cyrus, MN
- Walz International Space Drug State Corporation of Dakota, MN
- Walz International Space Drug State Corporation of Dalton, MN
- Walz International Space Drug State Corporation of Danube, MN
- Walz International Space Drug State Corporation of Darwin, MN
- Walz International Space Drug State Corporation of Dassel, MN
- Walz International Space Drug State Corporation of Dawson, MN
- Walz International Space Drug State Corporation of Dayton, MN
- Walz International Space Drug State Corporation of Deer Creek, MN
- Walz International Space Drug State Corporation of Deer River, MN
- Walz International Space Drug State Corporation of Deerwood, MN
- Walz International Space Drug State Corporation of Delano, MN
- Walz International Space Drug State Corporation of Delavan, MN
- Walz International Space Drug State Corporation of Detroit Lakes, MN
- Walz International Space Drug State Corporation of Dexter, MN
- Walz International Space Drug State Corporation of Dilworth, MN
- Walz International Space Drug State Corporation of Dodge Center, MN
- Walz International Space Drug State Corporation of Donnelly, MN
- Walz International Space Drug State Corporation of Dover, MN
- Walz International Space Drug State Corporation of Duluth, MN
- Walz International Space Drug State Corporation of Dundas, MN
- Walz International Space Drug State Corporation of Eagle Bend, MN
- Walz International Space Drug State Corporation of Eagle Lake, MN
- Walz International Space Drug State Corporation of East Grand Forks, MN
- Walz International Space Drug State Corporation of Easton, MN
- Walz International Space Drug State Corporation of Echo, MN
- Walz International Space Drug State Corporation of Eden Prairie, MN
- Walz International Space Drug State Corporation of Eden Valley, MN
- Walz International Space Drug State Corporation of Edgerton, MN

- Walz International Space Drug State Corporation of Eitzen, MN
- Walz International Space Drug State Corporation of Elbow Lake, MN
- Walz International Space Drug State Corporation of Elgin, MN
- Walz International Space Drug State Corporation of Elk River, MN
- Walz International Space Drug State Corporation of Elko, MN
- Walz International Space Drug State Corporation of Ellendale, MN
- Walz International Space Drug State Corporation of Ellsworth, MN
- Walz International Space Drug State Corporation of Elmore, MN
- Walz International Space Drug State Corporation of Ely, MN
- Walz International Space Drug State Corporation of Elysian, MN
- Walz International Space Drug State Corporation of Emily, MN
- Walz International Space Drug State Corporation of Emmons, MN
- Walz International Space Drug State Corporation of Erskine, MN
- Walz International Space Drug State Corporation of Evansville, MN
- Walz International Space Drug State Corporation of Eveleth, MN
- Walz International Space Drug State Corporation of Excelsior, MN
- Walz International Space Drug State Corporation of Eyota, MN
- Walz International Space Drug State Corporation of Fairfax, MN
- Walz International Space Drug State Corporation of Fairmont, MN
- Walz International Space Drug State Corporation of Faribault, MN
- Walz International Space Drug State Corporation of Farmington, MN
- Walz International Space Drug State Corporation of Felton, MN
- Walz International Space Drug State Corporation of Fergus Falls, MN
- Walz International Space Drug State Corporation of Fertile, MN
- Walz International Space Drug State Corporation of Fifty Lakes, MN
- Walz International Space Drug State Corporation of Finlayson, MN
- Walz International Space Drug State Corporation of Fisher, MN
- Walz International Space Drug State Corporation of Flensburg, MN
- Walz International Space Drug State Corporation of Floodwood, MN
- Walz International Space Drug State Corporation of Foley, MN
- Walz International Space Drug State Corporation of Forest Lake, MN
- Walz International Space Drug State Corporation of Foreston, MN
- Walz International Space Drug State Corporation of Fosston, MN

- Walz International Space Drug State Corporation of Fountain, MN
- Walz International Space Drug State Corporation of Foreston, MN
- Walz International Space Drug State Corporation of Fosston, MN
- Walz International Space Drug State Corporation of Fountain, MN
- Walz International Space Drug State Corporation of Franklin, MN
- Walz International Space Drug State Corporation of Frazee, MN
- Walz International Space Drug State Corporation of Freeborn, MN
- Walz International Space Drug State Corporation of Freeport, MN
- Walz International Space Drug State Corporation of Frost, MN
- Walz International Space Drug State Corporation of Fulda, MN
- Walz International Space Drug State Corporation of Garfield, MN
- Walz International Space Drug State Corporation of Garrison, MN
- Walz International Space Drug State Corporation of Gary, MN
- Walz International Space Drug State Corporation of Gaylor, MN
- Walz International Space Drug State Corporation of Geneva, MN
- Walz International Space Drug State Corporation of Ghent, MN
- Walz International Space Drug State Corporation of Gibbon, MN
- Walz International Space Drug State Corporation of Gilbert, MN
- Walz International Space Drug State Corporation of Gilman, MN
- Walz International Space Drug State Corporation of Glencoe, MN
- Walz International Space Drug State Corporation of Glenville, MN
- Walz International Space Drug State Corporation of Glenwood, MN
- Walz International Space Drug State Corporation of Glyndon, MN
- Walz International Space Drug State Corporation of Gonvick, MN
- Walz International Space Drug State Corporation of Good Thunder, MN
- Walz International Space Drug State Corporation of Goodhue, MN
- Walz International Space Drug State Corporation of Graceville, MN
- Walz International Space Drug State Corporation of Granada, MN
- Walz International Space Drug State Corporation of Grand Marais, MN
- Walz International Space Drug State Corporation of Grand Meadow, MN
- Walz International Space Drug State Corporation of Grand Rapids, MN
- Walz International Space Drug State Corporation of Granite Falls, MN
- Walz International Space Drug State Corporation of Green Isle, MN

- Walz International Space Drug State Corporation of Greenbush, MN
- Walz International Space Drug State Corporation of Greenwald, MN
- Walz International Space Drug State Corporation of Grey Eagle, MN
- Walz International Space Drug State Corporation of Grove City, MN
- Walz International Space Drug State Corporation of Grygla, MN
- Walz International Space Drug State Corporation of Hackensack, MN
- Walz International Space Drug State Corporation of Hallock, MN
- Walz International Space Drug State Corporation of Halstad, MN
- Walz International Space Drug State Corporation of Hamburg, MN
- Walz International Space Drug State Corporation of Hampton, MN
- Walz International Space Drug State Corporation of Hancock, MN
- Walz International Space Drug State Corporation of Hancock, MN
- Walz International Space Drug State Corporation of Hanley Falls, MN
- Walz International Space Drug State Corporation of Hanover, MN
- Walz International Space Drug State Corporation of Hanska, MN
- Walz International Space Drug State Corporation of Hardwick, MN
- Walz International Space Drug State Corporation of Harmony, MN
- Walz International Space Drug State Corporation of Harris, MN
- Walz International Space Drug State Corporation of Hartland, MN
- Walz International Space Drug State Corporation of Hastings, MN
- Walz International Space Drug State Corporation of Hawley, MN
- Walz International Space Drug State Corporation of Hayfield, MN
- Walz International Space Drug State Corporation of Hayward, MN
- Walz International Space Drug State Corporation of Hector, MN
- Walz International Space Drug State Corporation of Henderson, MN
- Walz International Space Drug State Corporation of Hendricks, MN
- Walz International Space Drug State Corporation of Hendrum, MN
- Walz International Space Drug State Corporation of Henning, MN
- Walz International Space Drug State Corporation of Herman, MN
- Walz International Space Drug State Corporation of Heron Lake, MN
- Walz International Space Drug State Corporation of Hewitt, MN
- Walz International Space Drug State Corporation of Hibbing, MN
- Walz International Space Drug State Corporation of Hill City, MN

- Walz International Space Drug State Corporation of Hills, MN
- Walz International Space Drug State Corporation of Hinckley, MN
- Walz International Space Drug State Corporation of Hitterdal, MN
- Walz International Space Drug State Corporation of Hoffman, MN
- Walz International Space Drug State Corporation of Hokah, MN
- Walz International Space Drug State Corporation of Holdingford, MN
- Walz International Space Drug State Corporation of Holland, MN
- Walz International Space Drug State Corporation of Hollandale, MN
- Walz International Space Drug State Corporation of Hopkins, MN
- Walz International Space Drug State Corporation of Houston, MN
- Walz International Space Drug State Corporation of Howard Lake, MN
- Walz International Space Drug State Corporation of Hoyt Lakes, MN
- Walz International Space Drug State Corporation of Hugo, MN
- Walz International Space Drug State Corporation of Hutchinson, MN
- Walz International Space Drug State Corporation of International Falls, MN
- Walz International Space Drug State Corporation of Inver Grove Heights, MN
- Walz International Space Drug State Corporation of Ironton, MN
- Walz International Space Drug State Corporation of Isanti, MN
- Walz International Space Drug State Corporation of Isle, MN
- Walz International Space Drug State Corporation of Ivanhoe, MN
- Walz International Space Drug State Corporation of Jackson, MN
- Walz International Space Drug State Corporation of Janesville, MN
- Walz International Space Drug State Corporation of Jasper, MN
- Walz International Space Drug State Corporation of Jeffers, MN
- Walz International Space Drug State Corporation of Jenkins, MN
- Walz International Space Drug State Corporation of Jordan, MN
- Walz International Space Drug State Corporation of Kandiyohi, MN
- Walz International Space Drug State Corporation of Karlstad, MN
- Walz International Space Drug State Corporation of Kasota, MN
- Walz International Space Drug State Corporation of Kasson, MN
- Walz International Space Drug State Corporation of Keewatin, MN
- Walz International Space Drug State Corporation of Kelliher, MN
- Walz International Space Drug State Corporation of Kellogg, MN

- Walz International Space Drug State Corporation of Kennedy, MN
- Walz International Space Drug State Corporation of Kensington, MN
- Walz International Space Drug State Corporation of Kenyon, MN
- Walz International Space Drug State Corporation of Kerkhoven, MN
- Walz International Space Drug State Corporation of Kiester, MN
- Walz International Space Drug State Corporation of Kimball, MN
- Walz International Space Drug State Corporation of La Crescent, MN
- Walz International Space Drug State Corporation of Lafayette, MN
- Walz International Space Drug State Corporation of Lake Benton, MN
- Walz International Space Drug State Corporation of Lake Bronson, MN
- Walz International Space Drug State Corporation of Lake City, MN
- Walz International Space Drug State Corporation of Lake Crystal, MN
- Walz International Space Drug State Corporation of Lake Elmo, MN
- Walz International Space Drug State Corporation of Lake Lillian, MN
- Walz International Space Drug State Corporation of Lake Park, MN
- Walz International Space Drug State Corporation of Lake Wilson, MN
- Walz International Space Drug State Corporation of Lakefield, MN
- Walz International Space Drug State Corporation of Lakeland, MN
- Walz International Space Drug State Corporation of Lakeville, MN
- Walz International Space Drug State Corporation of Lamberton, MN
- Walz International Space Drug State Corporation of Lancaster, MN
- Walz International Space Drug State Corporation of Lanesboro, MN
- Walz International Space Drug State Corporation of Le Center, MN
- Walz International Space Drug State Corporation of Le Roy, MN
- Walz International Space Drug State Corporation of Le Sueur, MN
- Walz International Space Drug State Corporation of Leota, MN
- Walz International Space Drug State Corporation of Lester Prairie, MN
- Walz International Space Drug State Corporation of Lewiston, MN
- Walz International Space Drug State Corporation of Lewisville, MN
- Walz International Space Drug State Corporation of Lindstrom, MN
- Walz International Space Drug State Corporation of Lismore, MN
- Walz International Space Drug State Corporation of Litchfield, MN
- Walz International Space Drug State Corporation of Little Falls, MN

- Walz International Space Drug State Corporation of Littlefork, MN
- Walz International Space Drug State Corporation of Long Lake, MN
- Walz International Space Drug State Corporation of Long Prairie, MN
- Walz International Space Drug State Corporation of Lonsdale, MN
- Walz International Space Drug State Corporation of Loretto, MN
- Walz International Space Drug State Corporation of Lowry, MN
- Walz International Space Drug State Corporation of Lucan, MN
- Walz International Space Drug State Corporation of Luverne, MN
- Walz International Space Drug State Corporation of Lyle, MN
- Walz International Space Drug State Corporation of Lynd, MN
- Walz International Space Drug State Corporation of Mabel, MN
- Walz International Space Drug State Corporation of Madelia, MN
- Walz International Space Drug State Corporation of Madison Lake, MN
- Walz International Space Drug State Corporation of Madison, MN
- Walz International Space Drug State Corporation of Magnolia, MN
- Walz International Space Drug State Corporation of Mahnomen, MN
- Walz International Space Drug State Corporation of Mankato, MN
- Walz International Space Drug State Corporation of Mantorville, MN
- Walz International Space Drug State Corporation of Maple Lake, MN
- Walz International Space Drug State Corporation of Maple Plain, MN
- Walz International Space Drug State Corporation of Mapleton, MN
- Walz International Space Drug State Corporation of Marble, MN
- Walz International Space Drug State Corporation of Marine On Saint Croix, MN
- Walz International Space Drug State Corporation of Marshall, MN
- Walz International Space Drug State Corporation of Mayer, MN
- Walz International Space Drug State Corporation of Maynard, MN
- Walz International Space Drug State Corporation of Mazeppa, MN
- Walz International Space Drug State Corporation of Mcgregor, MN
- Walz International Space Drug State Corporation of Mcintosh, MN
- Walz International Space Drug State Corporation of Medford, MN
- Walz International Space Drug State Corporation of Melrose, MN
- Walz International Space Drug State Corporation of Menahga, MN
- Walz International Space Drug State Corporation of Mendota, MN

- Walz International Space Drug State Corporation of Middle River, MN
- Walz International Space Drug State Corporation of Milaca, MN
- Walz International Space Drug State Corporation of Milan, MN
- Walz International Space Drug State Corporation of Milroy, MN
- Walz International Space Drug State Corporation of Miltona, MN
- Walz International Space Drug State Corporation of Minneapolis, MN
- Walz International Space Drug State Corporation of Minneota, MN
- Walz International Space Drug State Corporation of Minnesota City, MN
- Walz International Space Drug State Corporation of Minnesota Lake, MN
- Walz International Space Drug State Corporation of Minnetonka Beach, MN
- Walz International Space Drug State Corporation of Minnetonka, MN
- Walz International Space Drug State Corporation of Montevideo, MN
- Walz International Space Drug State Corporation of Montgomery, MN
- Walz International Space Drug State Corporation of Monticello, MN
- Walz International Space Drug State Corporation of Montrose, MN
- Walz International Space Drug State Corporation of Moorhead, MN
- Walz International Space Drug State Corporation of Moose Lake, MN
- Walz International Space Drug State Corporation of Mora, MN
- Walz International Space Drug State Corporation of Morgan, MN
- Walz International Space Drug State Corporation of Morris, MN
- Walz International Space Drug State Corporation of Morristown, MN
- Walz International Space Drug State Corporation of Morton, MN
- Walz International Space Drug State Corporation of Motley, MN
- Walz International Space Drug State Corporation of Mound, MN
- Walz International Space Drug State Corporation of Mountain Iron, MN
- Walz International Space Drug State Corporation of Mountain Lake, MN
- Walz International Space Drug State Corporation of Murdock, MN
- Walz International Space Drug State Corporation of Nashwauk, MN
- Walz International Space Drug State Corporation of Naytahwaush, MN
- Walz International Space Drug State Corporation of Nerstrand, MN
- Walz International Space Drug State Corporation of Nevis, MN
- Walz International Space Drug State Corporation of New Auburn, MN
- Walz International Space Drug State Corporation of New Germany, MN

- Walz International Space Drug State Corporation of New London, MN
- Walz International Space Drug State Corporation of New Market, MN
- Walz International Space Drug State Corporation of New Munich, MN
- Walz International Space Drug State Corporation of New Prague, MN
- Walz International Space Drug State Corporation of New Richland, MN
- Walz International Space Drug State Corporation of New Ulm, MN
- Walz International Space Drug State Corporation of New York Mills, MN
- Walz International Space Drug State Corporation of Newfolden, MN
- Walz International Space Drug State Corporation of Newport, MN
- Walz International Space Drug State Corporation of Nicollet, MN
- Walz International Space Drug State Corporation of Nisswa, MN
- Walz International Space Drug State Corporation of North Branch, MN
- Walz International Space Drug State Corporation of Northfield, MN
- Walz International Space Drug State Corporation of Northome, MN
- Walz International Space Drug State Corporation of Northrop, MN
- Walz International Space Drug State Corporation of Norwood, MN
- Walz International Space Drug State Corporation of Oak Park, MN
- Walz International Space Drug State Corporation of Ogilvie, MN
- Walz International Space Drug State Corporation of Oklee, MN
- Walz International Space Drug State Corporation of Olivia, MN
- Walz International Space Drug State Corporation of Onamia, MN
- Walz International Space Drug State Corporation of Oronoco, MN
- Walz International Space Drug State Corporation of Orr, MN
- Walz International Space Drug State Corporation of Ortonville, MN
- Walz International Space Drug State Corporation of Osakis, MN
- Walz International Space Drug State Corporation of Oslo, MN
- Walz International Space Drug State Corporation of Osseo, MN
- Walz International Space Drug State Corporation of Ostrander, MN
- Walz International Space Drug State Corporation of Ottertail, MN
- Walz International Space Drug State Corporation of Owatonna, MN
- Walz International Space Drug State Corporation of Park Rapids, MN
- Walz International Space Drug State Corporation of Parkers Prairie, MN
- Walz International Space Drug State Corporation of Paynesville, MN

- Walz International Space Drug State Corporation of Pelican Rapids, MN
- Walz International Space Drug State Corporation of Pemberton, MN
- Walz International Space Drug State Corporation of Pennock, MN
- Walz International Space Drug State Corporation of Pequot Lakes, MN
- Walz International Space Drug State Corporation of Perham, MN
- Walz International Space Drug State Corporation of Peterson, MN
- Walz International Space Drug State Corporation of Pierz, MN
- Walz International Space Drug State Corporation of Pillager, MN
- Walz International Space Drug State Corporation of Pine City, MN
- Walz International Space Drug State Corporation of Pine Island, MN
- Walz International Space Drug State Corporation of Pine River, MN
- Walz International Space Drug State Corporation of Pipestone, MN
- Walz International Space Drug State Corporation of Plainview, MN
- Walz International Space Drug State Corporation of Plato, MN
- Walz International Space Drug State Corporation of Plummer, MN
- Walz International Space Drug State Corporation of Ponemah, MN
- Walz International Space Drug State Corporation of Preston, MN
- Walz International Space Drug State Corporation of Princeton, MN
- Walz International Space Drug State Corporation of Prinsburg, MN
- Walz International Space Drug State Corporation of Prior Lake, MN
- Walz International Space Drug State Corporation of Racine, MN
- Walz International Space Drug State Corporation of Randall, MN
- Walz International Space Drug State Corporation of Randolph, MN
- Walz International Space Drug State Corporation of Raymond, MN
- Walz International Space Drug State Corporation of Red Lake Falls, MN
- Walz International Space Drug State Corporation of Red Wing, MN
- Walz International Space Drug State Corporation of Redby, MN
- Walz International Space Drug State Corporation of Redlake, MN
- Walz International Space Drug State Corporation of Redwood Falls, MN
- Walz International Space Drug State Corporation of Remer, MN
- Walz International Space Drug State Corporation of Renville, MN
- Walz International Space Drug State Corporation of Rice, MN
- Walz International Space Drug State Corporation of Richmond, MN

- Walz International Space Drug State Corporation of Rochester, MN
- Walz International Space Drug State Corporation of Rock Creek, MN
- Walz International Space Drug State Corporation of Rockford, MN
- Walz International Space Drug State Corporation of Rockville, MN
- Walz International Space Drug State Corporation of Rogers, MN
- Walz International Space Drug State Corporation of Rollingstone, MN
- Walz International Space Drug State Corporation of Rose Creek, MN
- Walz International Space Drug State Corporation of Roseau, MN
- Walz International Space Drug State Corporation of Rosemount, MN
- Walz International Space Drug State Corporation of Rothsay, MN
- Walz International Space Drug State Corporation of Round Lake, MN
- Walz International Space Drug State Corporation of Royalton, MN
- Walz International Space Drug State Corporation of Rush City, MN
- Walz International Space Drug State Corporation of Rushford, MN
- Walz International Space Drug State Corporation of Rushmore, MN
- Walz International Space Drug State Corporation of Russell, MN
- Walz International Space Drug State Corporation of Ruthton, MN
- Walz International Space Drug State Corporation of Sabin, MN
- Walz International Space Drug State Corporation of Sacred Heart, MN
- Walz International Space Drug State Corporation of Saint Bonifacius, MN
- Walz International Space Drug State Corporation of Saint Charles, MN
- Walz International Space Drug State Corporation of Saint Clair
- Walz International Space Drug State Corporation of Saint Cloud, MN
- Walz International Space Drug State Corporation of Saint Francis, MN
- Walz International Space Drug State Corporation of Saint Hilaire, MN
- Walz International Space Drug State Corporation of Saint James, MN
- Walz International Space Drug State Corporation of Saint Joseph, MN
- Walz International Space Drug State Corporation of Saint Martin, MN
- Walz International Space Drug State Corporation of Saint Michael, MN
- Walz International Space Drug State Corporation of Saint Paul Park, MN
- Walz International Space Drug State Corporation of Saint Paul, MN
- Walz International Space Drug State Corporation of Saint Peter, MN
- Walz International Space Drug State Corporation of Saint Stephen, MN

- Walz International Space Drug State Corporation of Sanborn, MN
- Walz International Space Drug State Corporation of Sandstone, MN
- Walz International Space Drug State Corporation of Sartell, MN
- Walz International Space Drug State Corporation of Sauk Centre, MN
- Walz International Space Drug State Corporation of Sauk Rapids, MN
- Walz International Space Drug State Corporation of Savage, MN
- Walz International Space Drug State Corporation of Sebeka, MN
- Walz International Space Drug State Corporation of Shafer, MN
- Walz International Space Drug State Corporation of Shakopee, MN
- Walz International Space Drug State Corporation of Shelly, MN
- Walz International Space Drug State Corporation of Sherburn, MN
- Walz International Space Drug State Corporation of Silver Bay, MN
- Walz International Space Drug State Corporation of Silver Lake, MN
- Walz International Space Drug State Corporation of Slayton, MN
- Walz International Space Drug State Corporation of Sleepy Eye, MN
- Walz International Space Drug State Corporation of South Haven, MN
- Walz International Space Drug State Corporation of South Saint Paul, MN
- Walz International Space Drug State Corporation of Spicer, MN
- Walz International Space Drug State Corporation of Spring Grove, MN
- Walz International Space Drug State Corporation of Spring Lake, MN
- Walz International Space Drug State Corporation of Spring Park, MN
- Walz International Space Drug State Corporation of Spring Valley, MN
- Walz International Space Drug State Corporation of Springfield, MN
- Walz International Space Drug State Corporation of Stacy, MN
- Walz International Space Drug State Corporation of Staples, MN
- Walz International Space Drug State Corporation of Starbuck, MN
- Walz International Space Drug State Corporation of Stephen, MN
- Walz International Space Drug State Corporation of Stewart, MN
- Walz International Space Drug State Corporation of Stewartville, MN
- Walz International Space Drug State Corporation of Stillwater, MN
- Walz International Space Drug State Corporation of Stockton, MN
- Walz International Space Drug State Corporation of Storden, MN
- Walz International Space Drug State Corporation of Sturgeon Lake, MN

- Walz International Space Drug State Corporation of Swanville, MN
- Walz International Space Drug State Corporation of Taconite, MN
- Walz International Space Drug State Corporation of Taunton, MN
- Walz International Space Drug State Corporation of Taylors Falls, MN
- Walz International Space Drug State Corporation of Thief River Falls, MN
- Walz International Space Drug State Corporation of Tower, MN
- Walz International Space Drug State Corporation of Tracy, MN
- Walz International Space Drug State Corporation of Trimont, MN
- Walz International Space Drug State Corporation of Truman, MN
- Walz International Space Drug State Corporation of Twin Valley, MN
- Walz International Space Drug State Corporation of Two Harbors, MN
- Walz International Space Drug State Corporation of Tyler, MN
- Walz International Space Drug State Corporation of Ulen, MN
- Walz International Space Drug State Corporation of Underwood, MN
- Walz International Space Drug State Corporation of Upsala, MN
- Walz International Space Drug State Corporation of Utica, MN
- Walz International Space Drug State Corporation of Vergas, MN
- Walz International Space Drug State Corporation of Vermillion, MN
- Walz International Space Drug State Corporation of Verndale, MN
- Walz International Space Drug State Corporation of Vernon Center, MN
- Walz International Space Drug State Corporation of Vesta, MN
- Walz International Space Drug State Corporation of Victoria, MN
- Walz International Space Drug State Corporation of Villard, MN
- Walz International Space Drug State Corporation of Virginia, MN
- Walz International Space Drug State Corporation of Wabasha, MN
- Walz International Space Drug State Corporation of Wabasso, MN
- Walz International Space Drug State Corporation of Waconia, MN
- Walz International Space Drug State Corporation of Wadena, MN
- Walz International Space Drug State Corporation of Wahkon, MN
- Walz International Space Drug State Corporation of Waite Park, MN
- Walz International Space Drug State Corporation of Waldorf, MN
- Walz International Space Drug State Corporation of Walker, MN
- Walz International Space Drug State Corporation of Walnut Grove, MN

- Walz International Space Drug State Corporation of Wanamingo, MN
- Walz International Space Drug State Corporation of Warren, MN
- Walz International Space Drug State Corporation of Warroad, MN
- Walz International Space Drug State Corporation of Waseca, MN
- Walz International Space Drug State Corporation of Watertown, MN
- Walz International Space Drug State Corporation of Waterville, MN
- Walz International Space Drug State Corporation of Watkins, MN
- Walz International Space Drug State Corporation of Watson, MN
- Walz International Space Drug State Corporation of Waubun, MN
- Walz International Space Drug State Corporation of Waverly, MN
- Walz International Space Drug State Corporation of Wayzata, MN
- Walz International Space Drug State Corporation of Welcome, MN
- Walz International Space Drug State Corporation of Wells, MN
- Walz International Space Drug State Corporation of West Concord, MN
- Walz International Space Drug State Corporation of Westbrook, MN
- Walz International Space Drug State Corporation of Wheaton, MN
- Walz International Space Drug State Corporation of White Earth, MN
- Walz International Space Drug State Corporation of Willernie, MN
- Walz International Space Drug State Corporation of Williams, MN
- Walz International Space Drug State Corporation of Willmar, MN
- Walz International Space Drug State Corporation of Willow River, MN
- Walz International Space Drug State Corporation of Wilmont, MN
- Walz International Space Drug State Corporation of Windom, MN
- Walz International Space Drug State Corporation of Winger, MN
- Walz International Space Drug State Corporation of Winnebago, MN
- Walz International Space Drug State Corporation of Winona, MN
- Walz International Space Drug State Corporation of Winsted, MN
- Walz International Space Drug State Corporation of Winthrop, MN
- Walz International Space Drug State Corporation of Wood Lake, MN
- Walz International Space Drug State Corporation of Worthington, MN
- Walz International Space Drug State Corporation of Wrenshall, MN
- Walz International Space Drug State Corporation of Wykoff, MN
- Walz International Space Drug State Corporation of Wyoming, MN

- **Walz International Space Drug State Corporation of Zimmerman, MN**
- **Walz International Space Drug State Corporation of Zumbrota, MN**

There are a total of 912 cities in Minnesota. By selling containers to each city through this corporation, the projected profit sales amount to $54.7 trillion annually. To fulfill the Minnesota market, we need 52 International Space State Corporations.

- People might think it's unconventional, but...
- We are the people who live in the real world and see it happening, even if it's illegal.
- We are the people who believe that directing this money to the proper use can bring free healthcare, education, and other benefits to the people.
- We are the people who believe that, through this, we can contribute to space research and colonization of planets.
- We are the people who believe in the prosperity of countries and the world.

To develop and present this project to the public, we need:

1. Financing
2. Selection of professional employees, specialists
3. Selection and purchase of special materials, technologies for development
4. Open government facilities for public access

Revenue from the sales will be directed towards healthcare, military agency expenses, and the space project program created by Georgiy S Garbuz. This program aims to build a national Garbuz Space Academy and national Garbuz Space Academy schools worldwide to grant foster care and other kids access to education. Funding for grants to educate these kids will be derived from directed taxes on Georgiy S Garbuz's science works and all projects.

The project was initially conceived by Georgiy Sergeyevich Garbuz in 2007 and is being developed under the name "Project For The Future" (G.S. Garbuz) at the National Garbuz Space Academy Trust Corporation

DANIELKA INTERNATIONAL SPACE DRUG STATE CORPORATION OF NEW YORK

The International Intergalactic State Corporation is designed to boost the economy of countries, states, and the world. Its purpose is to explore space across galaxies. Thanks to the president, with our assistance, non-addictive drugs became legal in the United States. We aim to legalize marijuana worldwide. The corporation is designed to purchase and grow marijuana in economically challenged countries.

According to 2003 prices in cities like Esik (Kazakhstan), Novaia Sinjereia (Moldova), and Chernovchi (Ukraine), the price for 1 kilogram of marijuana was $5-10. In the United States, the same kilogram cost $3000. This proves to be a highly profitable business in the black market. According to our research, one container sold to a city each week would bring in profit sales of $59.6 billion annually. To fulfill the state of New York, we need 63 International State Corporations, bringing in a profit sale of $3.76 trillion. To fulfill the entire USA market, we need 2.37 million containers, each weighing 40,000 pounds.

Every state controls the drug market in a country through state corporations, allowing people to open partner stores in every city. As we transition from non-legal to legal sales in the country, a new government project program has been initiated to legalize former drug dealers in legal businesses. A special program has been created, which needs acceptance from the people to facilitate work in a legal business and the legal sale of marijuana in the country.

Danielka International Space Drug State Corporation of New York

- 40%: State Government
- 20%: Imperial Space Federation (USA Government)
- 10%: Federal Government
- 10%: NASA
- 5%: CIA, Interpol, FBI, or Military
- 5%: State Police or Federal Police
- 5%: Israel, Ukraine, Kazakhstan, or any other country government
- 5%: Boris, Local Businessmen, or Former Drug Dealer (subject to meeting all government agency requirements)

Danielka International Space Drug State Corporation of Accord, NY

- 40%: Danielka International Space Drug State Corporation of New York (Owner)
- 10%: Imperial Space Federation
- 50%: Local Businessmen or Former Drug Dealer (subject to meeting all government agency requirements)

Danielka International Space Drug State Corporation of Adams Center, NY

- 40%: Danielka International Space Drug State Corporation of New York (Owner)
- 10%: Imperial Space Federation
- 50%: Local Businessmen or Former Drug Dealer (subject to meeting all government agency requirements)

Danielka International Space Drug State Corporation of Adams, NY

- 40%: Danielka International Space Drug State Corporation of New York (Owner)
- 10%: Imperial Space Federation
- 50%: Local Businessmen or Former Drug Dealer (subject to meeting all government agency requirements)

Danielka International Space Drug State Corporation of Addison, NY

- 40%: Danielka International Space Drug State Corporation of New York (Owner)
- 10%: Imperial Space Federation
- 50%: Local Businessmen or Former Drug Dealer (subject to meeting all government agency requirements)

Danielka International Space Drug State Corporation of Akron, NY

- 40%: Danielka International Space Drug State Corporation of New York (Owner)
- 10%: Imperial Space Federation
- 50%: Local Businessmen or Former Drug Dealer (subject to meeting all government agency requirements)

Danielka International Space Drug State Corporation of Albany, NY

- 40%: Danielka International Space Drug State Corporation of New York (Owner)
- 10%: Imperial Space Federation
- 50%: Local Businessmen or Former Drug Dealer (subject to meeting all government agency requirements)

Danielka International Space Drug State Corporation of Albertson, NY

- 40%: Danielka International Space Drug State Corporation of New York (Owner)
- 10%: Imperial Space Federation
- 50%: Local Businessmen or Former Drug Dealer (subject to meeting all government agency requirements)

Danielka International Space Drug State Corporation of Albion, NY

- 40%: Danielka International Space Drug State Corporation of New York (Owner)
- 10%: Imperial Space Federation
- 50%: Local Businessmen or Former Drug Dealer (subject to meeting all government agency requirements)

Danielka International Space Drug State Corporation of Alden, NY

- 40%: Danielka International Space Drug State Corporation of New York (Owner)
- 10%: Imperial Space Federation
- 50%: Local Businessmen or Former Drug Dealer (subject to meeting all government agency requirements)

Danielka International Space Drug State Corporation of Alexander, NY

- 40%: Danielka International Space Drug State Corporation of New York (Owner)
- 10%: Imperial Space Federation
- 50%: Local Businessmen or Former Drug Dealer (subject to meeting all government agency requirements)

Danielka International Space Drug State Corporation of Alexandria Bay, NY

- 40%: Danielka International Space Drug State Corporation of New York (Owner)
- 10%: Imperial Space Federation
- 50%: Local Businessmen or Former Drug Dealer (subject to meeting all government agency requirements)

Danielka International Space Drug State Corporation of Alfred, NY

- 40%: Danielka International Space Drug State Corporation of New York (Owner)
- 10%: Imperial Space Federation
- 50%: Local Businessmen or Former Drug Dealer (subject to meeting all government agency requirements)

Danielka International Space Drug State Corporation of Allegany, NY

- 40%: Danielka International Space Drug State Corporation of New York (Owner)
- 10%: Imperial Space Federation
- 50%: Local Businessmen or Former Drug Dealer (subject to meeting all government agency requirements)

Danielka International Space Drug State Corporation of Allentown, NY

- 40%: Danielka International Space Drug State Corporation of New York (Owner)
- 10%: Imperial Space Federation
- 50%: Local Businessmen or Former Drug Dealer (subject to meeting all government agency requirements)

Danielka International Space Drug State Corporation of Alma, NY

- 40%: Danielka International Space Drug State Corporation of New York (Owner)
- 10%: Imperial Space Federation
- 50%: Local Businessmen or Former Drug Dealer (subject to meeting all government agency requirements)

Danielka International Space Drug State Corporation of Almond, NY

- 40%: Danielka International Space Drug State Corporation of New York (Owner)
- 10%: Imperial Space Federation
- 50%: Local Businessmen or Former Drug Dealer (subject to meeting all government agency requirements)

Danielka International Space Drug State Corporation of Altamont, NY

- 40%: Danielka International Space Drug State Corporation of New York (Owner)
- 10%: Imperial Space Federation
- 50%: Local Businessmen or Former Drug Dealer (subject to meeting all government agency requirements)

Danielka International Space Drug State Corporation of Atltmar, NY

- 40%: Danielka International Space Drug State Corporation of New York (Owner)
- 10%: Imperial Space Federation
- 50%: Local Businessmen or Former Drug Dealer (subject to meeting all government agency requirements)

Danielka International Space Drug State Corporation of Altona, NY

- 40%: Danielka International Space Drug State Corporation of New York (Owner)
- 10%: Imperial Space Federation
- 50%: Local Businessmen or Former Drug Dealer (subject to meeting all government agency requirements)

Danielka International Space Drug State Corporation of Amagansett, NY

- 40%: Danielka International Space Drug State Corporation of New York (Owner)
- 10%: Imperial Space Federation
- 50%: Local Businessmen or Former Drug Dealer (subject to meeting all government agency requirements)

Danielka International Space Drug State Corporation of Amenia, NY

- 40%: Danielka International Space Drug State Corporation of New York (Owner)
- 10%: Imperial Space Federation
- 50%: Local Businessmen or Former Drug Dealer (subject to meeting all government agency requirements)

Danielka International Space Drug State Corporation of Amityville, NY

- 40%: Danielka International Space Drug State Corporation of New York (Owner)
- 10%: Imperial Space Federation
- 50%: Local Businessmen or Former Drug Dealer (subject to meeting all government agency requirements)

Danielka International Space Drug State Corporation of Amsterdam, NY

- 40%: Danielka International Space Drug State Corporation of New York (Owner)
- 10%: Imperial Space Federation
- 50%: Local Businessmen or Former Drug Dealer (subject to meeting all government agency requirements)

Danielka International Space Drug State Corporation of Ancram, NY

- 40%: Danielka International Space Drug State Corporation of New York (Owner)
- 10%: Imperial Space Federation
- 50%: Local Businessmen or Former Drug Dealer (subject to meeting all government agency requirements)

Danielka International Space Drug State Corporation of Andes, NY

- 40%: Danielka International Space Drug State Corporation of New York (Owner)
- 10%: Imperial Space Federation
- 50%: Local Businessmen or Former Drug Dealer (subject to meeting all government agency requirements)

In addition, Danielka International Space Drug State Corporations will be created for the other following cities in the state of Minnesota, operating with the same ownership structure as above:

- **Danielka International Space Drug State Corporation of Andover, NY**
- **Danielka International Space Drug State Corporation of Angelica, NY**
- **Danielka International Space Drug State Corporation of Angola, NY**
- **Danielka International Space Drug State Corporation of Antwerp, NY**
- **Danielka International Space Drug State Corporation of Apalachin, NY**
- **Danielka International Space Drug State Corporation of Aquebogue, NY**
- **Danielka International Space Drug State Corporation of Arcade, NY**
- **Danielka International Space Drug State Corporation of Ardsley, NY**

- Danielka International Space Drug State Corporation of Argyle, NY
- Danielka International Space Drug State Corporation of Arkport, NY
- Danielka International Space Drug State Corporation of Armonk, NY
- Danielka International Space Drug State Corporation of Ashland, NY
- Danielka International Space Drug State Corporation of Athens, NY
- Danielka International Space Drug State Corporation of Atlantic Beach, NY
- Danielka International Space Drug State Corporation of Attica, NY
- Danielka International Space Drug State Corporation of Au Sable Forks, NY
- Danielka International Space Drug State Corporation of Auburn, NY
- Danielka International Space Drug State Corporation of Aurora, NY
- Danielka International Space Drug State Corporation of Austerlitz, NY
- Danielka International Space Drug State Corporation of Ava, NY
- Danielka International Space Drug State Corporation of Averill Park, NY
- Danielka International Space Drug State Corporation of Avoca, NY
- Danielka International Space Drug State Corporation of Avon, NY
- Danielka International Space Drug State Corporation of Babylon, NY
- Danielka International Space Drug State Corporation of Bainbridge, NY
- Danielka International Space Drug State Corporation of Baldwin, NY
- Danielka International Space Drug State Corporation of Baldwinsville, NY
- Danielka International Space Drug State Corporation of Ballston Lake, NY
- Danielka International Space Drug State Corporation of Ballston Spa, NY
- Danielka International Space Drug State Corporation of Barker, NY
- Danielka International Space Drug State Corporation of Barneveld, NY
- Danielka International Space Drug State Corporation of Barton, NY
- Danielka International Space Drug State Corporation of Batavia, NY
- Danielka International Space Drug State Corporation of Bath, NY
- Danielka International Space Drug State Corporation of Bay Shore, NY
- Danielka International Space Drug State Corporation of Bayport, NY
- Danielka International Space Drug State Corporation of Bayville, NY
- Danielka International Space Drug State Corporation of Beacon, NY
- Danielka International Space Drug State Corporation of Bedford, NY
- Danielka International Space Drug State Corporation of Belfast, NY
- Danielka International Space Drug State Corporation of Bellerose, NY

- Danielka International Space Drug State Corporation of Bellmore, NY
- Danielka International Space Drug State Corporation of Bellport, NY
- Danielka International Space Drug State Corporation of Belmont, NY
- Danielka International Space Drug State Corporation of Bemus Point, NY
- Danielka International Space Drug State Corporation of Bergen, NY
- Danielka International Space Drug State Corporation of Berkshire, NY
- Danielka International Space Drug State Corporation of Berlin, NY
- Danielka International Space Drug State Corporation of Berne, NY
- Danielka International Space Drug State Corporation of Bethel, NY
- Danielka International Space Drug State Corporation of Bethpage, NY
- Danielka International Space Drug State Corporation of Big Flats, NY
- Danielka International Space Drug State Corporation of Binghamton, NY
- Danielka International Space Drug State Corporation of Black River, NY
- Danielka International Space Drug State Corporation of Blauvelt, NY
- Danielka International Space Drug State Corporation of Bloomfield, NY
- Danielka International Space Drug State Corporation of Blooming Grove, NY
- Danielka International Space Drug State Corporation of Bloomingburg, NY
- Danielka International Space Drug State Corporation of Blue Point, NY
- Danielka International Space Drug State Corporation of Bohemia, NY
- Danielka International Space Drug State Corporation of Bolivar, NY
- Danielka International Space Drug State Corporation of Bolton Landing, NY
- Danielka International Space Drug State Corporation of Bombay, NY
- Danielka International Space Drug State Corporation of Boonville, NY
- Danielka International Space Drug State Corporation of Boston, NY
- Danielka International Space Drug State Corporation of Bovina Center, NY
- Danielka International Space Drug State Corporation of Bradford, NY
- Danielka International Space Drug State Corporation of Brant, NY
- Danielka International Space Drug State Corporation of Brasher Falls, NY
- Danielka International Space Drug State Corporation of Brentwood, NY
- Danielka International Space Drug State Corporation of Brewerton, NY
- Danielka International Space Drug State Corporation of Briarcliff Manor, NY
- Danielka International Space Drug State Corporation of Bridgehampton, NY
- Danielka International Space Drug State Corporation of Bridgeport, NY

- Danielka International Space Drug State Corporation of Bridgewater, NY
- Danielka International Space Drug State Corporation of Brightwaters, NY
- Danielka International Space Drug State Corporation of Broadalbin, NY
- Danielka International Space Drug State Corporation of Brockport, NY
- Danielka International Space Drug State Corporation of Brocton, NY
- Danielka International Space Drug State Corporation of Bronxville, NY
- Danielka International Space Drug State Corporation of Brookfield, NY
- Danielka International Space Drug State Corporation of Brookhaven, NY
- Danielka International Space Drug State Corporation of Brownville, NY
- Danielka International Space Drug State Corporation of Brushton, NY
- Danielka International Space Drug State Corporation of Buchanan, NY
- Danielka International Space Drug State Corporation of Buffalo, NY
- Danielka International Space Drug State Corporation of Burdett, NY
- Danielka International Space Drug State Corporation of Burke, NY
- Danielka International Space Drug State Corporation of Burlington Flats, NY
- Danielka International Space Drug State Corporation of Byron, NY
- Danielka International Space Drug State Corporation of Cairo, NY
- Danielka International Space Drug State Corporation of Calcium, NY
- Danielka International Space Drug State Corporation of Caledonia, NY
- Danielka International Space Drug State Corporation of Caledonia, NY
- Danielka International Space Drug State Corporation of Callicoon, NY
- Danielka International Space Drug State Corporation of Calverton, NY
- Danielka International Space Drug State Corporation of Cambria Heights, NY
- Danielka International Space Drug State Corporation of Cambridge, NY
- Danielka International Space Drug State Corporation of Camden, NY
- Danielka International Space Drug State Corporation of Cameron, NY
- Danielka International Space Drug State Corporation of Camillus, NY
- Danielka International Space Drug State Corporation of Campbell, NY
- Danielka International Space Drug State Corporation of Canaan, NY
- Danielka International Space Drug State Corporation of Canajoharie, NY
- Danielka International Space Drug State Corporation of Canandaigua, NY
- Danielka International Space Drug State Corporation of Canaseraga, NY
- Danielka International Space Drug State Corporation of Canastota, NY

- Danielka International Space Drug State Corporation of Candor, NY
- Danielka International Space Drug State Corporation of Caneadea, NY
- Danielka International Space Drug State Corporation of Canisteo, NY
- Danielka International Space Drug State Corporation of Canton, NY
- Danielka International Space Drug State Corporation of Cape Vincent, NY
- Danielka International Space Drug State Corporation of Carle Place, NY
- Danielka International Space Drug State Corporation of Carlisle, NY
- Danielka International Space Drug State Corporation of Carmel, NY
- Danielka International Space Drug State Corporation of Caroga Lake, NY
- Danielka International Space Drug State Corporation of Carthage, NY
- Danielka International Space Drug State Corporation of Cassadaga, NY
- Danielka International Space Drug State Corporation of Castile, NY
- Danielka International Space Drug State Corporation of Castleton On Hudson, NY
- Danielka International Space Drug State Corporation of Castorland, NY
- Danielka International Space Drug State Corporation of Cato, NY
- Danielka International Space Drug State Corporation of Catskill, NY
- Danielka International Space Drug State Corporation of Cattaraugus, NY
- Danielka International Space Drug State Corporation of Cayuga, NY
- Danielka International Space Drug State Corporation of Cayuta, NY
- Danielka International Space Drug State Corporation of Cazenovia, NY
- Danielka International Space Drug State Corporation of Cedarhurst, NY
- Danielka International Space Drug State Corporation of Celoron, NY
- Danielka International Space Drug State Corporation of Center Moriches, NY
- Danielka International Space Drug State Corporation of Centereach, NY
- Danielka International Space Drug State Corporation of Centerport, NY
- Danielka International Space Drug State Corporation of Centerville, NY
- Danielka International Space Drug State Corporation of Central Islip, NY
- Danielka International Space Drug State Corporation of Central Square, NY
- Danielka International Space Drug State Corporation of Central Valley, NY
- Danielka International Space Drug State Corporation of Champlain, NY
- Danielka International Space Drug State Corporation of Chappaqua, NY
- Danielka International Space Drug State Corporation of Charlotteville, NY
- Danielka International Space Drug State Corporation of Chateaugay, NY

- Danielka International Space Drug State Corporation of Chatham, NY
- Danielka International Space Drug State Corporation of Chaumont, NY
- Danielka International Space Drug State Corporation of Chautauqua, NY
- Danielka International Space Drug State Corporation of Chazy, NY
- Danielka International Space Drug State Corporation of Chemung, NY
- Danielka International Space Drug State Corporation of Chenango Bridge, NY
- Danielka International Space Drug State Corporation of Chenango Forks, NY
- Danielka International Space Drug State Corporation of Cherry Creek, NY
- Danielka International Space Drug State Corporation of Cherry Valley, NY
- Danielka International Space Drug State Corporation of Chester, NY
- Danielka International Space Drug State Corporation of Chittenango, NY
- Danielka International Space Drug State Corporation of Churchville, NY
- Danielka International Space Drug State Corporation of Cicero, NY
- Danielka International Space Drug State Corporation of Cincinnatus, NY
- Danielka International Space Drug State Corporation of Clarence Center, NY
- Danielka International Space Drug State Corporation of Clarence, NY
- Danielka International Space Drug State Corporation of Clarendon, NY
- Danielka International Space Drug State Corporation of Clark Mills, NY
- Danielka International Space Drug State Corporation of Clarkson, NY
- Danielka International Space Drug State Corporation of Clarksville, NY
- Danielka International Space Drug State Corporation of Claverack, NY
- Danielka International Space Drug State Corporation of Clay, NY
- Danielka International Space Drug State Corporation of Clayton, NY
- Danielka International Space Drug State Corporation of Clayville, NY
- Danielka International Space Drug State Corporation of Cleveland, NY
- Danielka International Space Drug State Corporation of Clifton Park, NY
- Danielka International Space Drug State Corporation of Clifton Springs, NY
- Danielka International Space Drug State Corporation of Clinton, NY
- Danielka International Space Drug State Corporation of Clintondale, NY
- Danielka International Space Drug State Corporation of Clyde, NY
- Danielka International Space Drug State Corporation of Clymer, NY
- Danielka International Space Drug State Corporation of Cobleskill, NY
- Danielka International Space Drug State Corporation of Cochecton, NY

- Danielka International Space Drug State Corporation of Coeymans, NY
- Danielka International Space Drug State Corporation of Cohocton, NY
- Danielka International Space Drug State Corporation of Cohoes, NY
- Danielka International Space Drug State Corporation of Cold Brook, NY
- Danielka International Space Drug State Corporation of Cold Spring Harbor, NY
- Danielka International Space Drug State Corporation of Cold Spring, NY
- Danielka International Space Drug State Corporation of Colden, NY
- Danielka International Space Drug State Corporation of Collins, NY
- Danielka International Space Drug State Corporation of Colton, NY
- Danielka International Space Drug State Corporation of Commack, NY
- Danielka International Space Drug State Corporation of Conesus, NY
- Danielka International Space Drug State Corporation of Conewango Valley, NY
- Danielka International Space Drug State Corporation of Congers, NY
- Danielka International Space Drug State Corporation of Conklin, NY
- Danielka International Space Drug State Corporation of Constable, NY
- Danielka International Space Drug State Corporation of Constableville, NY
- Danielka International Space Drug State Corporation of Constantia, NY
- Danielka International Space Drug State Corporation of Cooperstown, NY
- Danielka International Space Drug State Corporation of Copake, NY
- Danielka International Space Drug State Corporation of Copenhagen, NY
- Danielka International Space Drug State Corporation of Copiague, NY
- Danielka International Space Drug State Corporation of Coram, NY
- Danielka International Space Drug State Corporation of Corfu, NY
- Danielka International Space Drug State Corporation of Corinth, NY
- Danielka International Space Drug State Corporation of Corning, NY
- Danielka International Space Drug State Corporation of Cornwall On Hudson, NY
- Danielka International Space Drug State Corporation of Cornwall, NY
- Danielka International Space Drug State Corporation of Cortland, NY
- Danielka International Space Drug State Corporation of Cortlandt Manor, NY
- Danielka International Space Drug State Corporation of Coxsackie, NY
- Danielka International Space Drug State Corporation of Cragsmoor, NY
- Danielka International Space Drug State Corporation of Croghan, NY
- Danielka International Space Drug State Corporation of Crompond, NY

- Danielka International Space Drug State Corporation of Croton On Hudson, NY
- Danielka International Space Drug State Corporation of Crown Point, NY
- Danielka International Space Drug State Corporation of Cuba, NY
- Danielka International Space Drug State Corporation of Cutchogue, NY
- Danielka International Space Drug State Corporation of Dannemora, NY
- Danielka International Space Drug State Corporation of Dansville, NY
- Danielka International Space Drug State Corporation of Darien Center, NY
- Danielka International Space Drug State Corporation of Davenport, NY
- Danielka International Space Drug State Corporation of Dayton, NY
- Danielka International Space Drug State Corporation of De Kalb Junction, NY
- Danielka International Space Drug State Corporation of De Peyster, NY
- Danielka International Space Drug State Corporation of De Ruyter, NY
- Danielka International Space Drug State Corporation of Deer Park, NY
- Danielka International Space Drug State Corporation of Deferiet, NY
- Danielka International Space Drug State Corporation of Delanson, NY
- Danielka International Space Drug State Corporation of Delevan, NY
- Danielka International Space Drug State Corporation of Delhi, NY
- Danielka International Space Drug State Corporation of Delmar, NY
- Danielka International Space Drug State Corporation of Denmark, NY
- Danielka International Space Drug State Corporation of Depauville, NY
- Danielka International Space Drug State Corporation of Depew, NY
- Danielka International Space Drug State Corporation of Deposit, NY
- Danielka International Space Drug State Corporation of Dexter, NY
- Danielka International Space Drug State Corporation of Dickinson Center, NY
- Danielka International Space Drug State Corporation of Dobbs Ferry, NY
- Danielka International Space Drug State Corporation of Dolgeville, NY
- Danielka International Space Drug State Corporation of Dover Plains, NY
- Danielka International Space Drug State Corporation of Dresden, NY
- Danielka International Space Drug State Corporation of Dryden, NY
- Danielka International Space Drug State Corporation of Duanesburg, NY
- Danielka International Space Drug State Corporation of Dundee, NY
- Danielka International Space Drug State Corporation of Dunkirk, NY
- Danielka International Space Drug State Corporation of Durham, NY

- Danielka International Space Drug State Corporation of Eagle Bay, NY
- Danielka International Space Drug State Corporation of Eagle Bridge, NY
- Danielka International Space Drug State Corporation of Earlville, NY
- Danielka International Space Drug State Corporation of East Aurora, NY
- Danielka International Space Drug State Corporation of East Bloomfield, NY
- Danielka International Space Drug State Corporation of East Greenbush, NY
- Danielka International Space Drug State Corporation of East Hampton, NY
- Danielka International Space Drug State Corporation of East Islip, NY
- Danielka International Space Drug State Corporation of East Marion, NY
- Danielka International Space Drug State Corporation of East Meadow, NY
- Danielka International Space Drug State Corporation of East Moriches, NY
- Danielka International Space Drug State Corporation of East Nassau, NY
- Danielka International Space Drug State Corporation of East Northport, NY
- Danielka International Space Drug State Corporation of East Norwich, NY
- Danielka International Space Drug State Corporation of East Otto, NY
- Danielka International Space Drug State Corporation of East Quogue, NY
- Danielka International Space Drug State Corporation of East Randolph, NY
- Danielka International Space Drug State Corporation of East Rochester, NY
- Danielka International Space Drug State Corporation of East Rockaway, NY
- Danielka International Space Drug State Corporation of East Syracuse, NY
- Danielka International Space Drug State Corporation of East Williamson, NY
- Danielka International Space Drug State Corporation of Eastchester, NY
- Danielka International Space Drug State Corporation of Eastport, NY
- Danielka International Space Drug State Corporation of Eaton, NY
- Danielka International Space Drug State Corporation of Eden, NY
- Danielka International Space Drug State Corporation of Edmeston, NY
- Danielka International Space Drug State Corporation of Edwards, NY
- Danielka International Space Drug State Corporation of Elba, NY
- Danielka International Space Drug State Corporation of Elbridge, NY
- Danielka International Space Drug State Corporation of Elizabethtown, NY
- Danielka International Space Drug State Corporation of Ellenburg, NY
- Danielka International Space Drug State Corporation of Ellenville, NY
- Danielka International Space Drug State Corporation of Ellicottville, NY

- Danielka International Space Drug State Corporation of Ellington, NY
- Danielka International Space Drug State Corporation of Ellisburg, NY
- Danielka International Space Drug State Corporation of Elma, NY
- Danielka International Space Drug State Corporation of Elmira, NY
- Danielka International Space Drug State Corporation of Elmont, NY
- Danielka International Space Drug State Corporation of Elmsford, NY
- Danielka International Space Drug State Corporation of Endicott, NY
- Danielka International Space Drug State Corporation of Endwell, NY
- Danielka International Space Drug State Corporation of Erin, NY
- Danielka International Space Drug State Corporation of Esopus, NY
- Danielka International Space Drug State Corporation of Esperance, NY
- Danielka International Space Drug State Corporation of Essex, NY
- Danielka International Space Drug State Corporation of Evans Mills, NY
- Danielka International Space Drug State Corporation of Fabius, NY
- Danielka International Space Drug State Corporation of Fair Haven, NY
- Danielka International Space Drug State Corporation of Fairport, NY
- Danielka International Space Drug State Corporation of Falconer, NY
- Danielka International Space Drug State Corporation of Fallsburg, NY
- Danielka International Space Drug State Corporation of Farmingdale, NY
- Danielka International Space Drug State Corporation of Farmington, NY
- Danielka International Space Drug State Corporation of Farmingville, NY
- Danielka International Space Drug State Corporation of Farnham, NY
- Danielka International Space Drug State Corporation of Fayette, NY
- Danielka International Space Drug State Corporation of Fayetteville, NY
- Danielka International Space Drug State Corporation of Fine, NY
- Danielka International Space Drug State Corporation of Fishers Island, NY
- Danielka International Space Drug State Corporation of Fishkill, NY
- Danielka International Space Drug State Corporation of Fleischmanns, NY
- Danielka International Space Drug State Corporation of Floral Park, NY
- Danielka International Space Drug State Corporation of Florida, NY
- Danielka International Space Drug State Corporation of Fonda, NY
- Danielka International Space Drug State Corporation of Forestburgh, NY
- Danielka International Space Drug State Corporation of Forestport, NY

- Danielka International Space Drug State Corporation of Forestville, NY
- Danielka International Space Drug State Corporation of Fort Ann, NY
- Danielka International Space Drug State Corporation of Fort Covington, NY
- Danielka International Space Drug State Corporation of Fort Drum, NY
- Danielka International Space Drug State Corporation of Fort Edward, NY
- Danielka International Space Drug State Corporation of Fort Johnson, NY
- Danielka International Space Drug State Corporation of Fort Montgomery, NY
- Danielka International Space Drug State Corporation of Fort Plain, NY
- Danielka International Space Drug State Corporation of Frankfort, NY
- Danielka International Space Drug State Corporation of Franklin Square, NY
- Danielka International Space Drug State Corporation of Franklin, NY
- Danielka International Space Drug State Corporation of Franklinville, NY
- Danielka International Space Drug State Corporation of Fredonia, NY
- Danielka International Space Drug State Corporation of Freedom, NY
- Danielka International Space Drug State Corporation of Freeport, NY
- Danielka International Space Drug State Corporation of Freeville, NY
- Danielka International Space Drug State Corporation of Fremont Center, NY
- Danielka International Space Drug State Corporation of Frewsburg, NY
- Danielka International Space Drug State Corporation of Friendship, NY
- Danielka International Space Drug State Corporation of Fulton, NY
- Danielka International Space Drug State Corporation of Fultonville, NY
- Danielka International Space Drug State Corporation of Gainesville, NY
- Danielka International Space Drug State Corporation of Galway, NY
- Danielka International Space Drug State Corporation of Garden City, NY
- Danielka International Space Drug State Corporation of Gardiner, NY
- Danielka International Space Drug State Corporation of Gasport, NY
- Danielka International Space Drug State Corporation of Geneseo, NY
- Danielka International Space Drug State Corporation of Geneva, NY
- Danielka International Space Drug State Corporation of Genoa, NY
- Danielka International Space Drug State Corporation of Georgetown, NY
- Danielka International Space Drug State Corporation of Germantown, NY
- Danielka International Space Drug State Corporation of Gerry, NY
- Danielka International Space Drug State Corporation of Ghent, NY

- Danielka International Space Drug State Corporation of Gilbertsville, NY
- Danielka International Space Drug State Corporation of Gilboa, NY
- Danielka International Space Drug State Corporation of Glasco, NY
- Danielka International Space Drug State Corporation of Glen Cove, NY
- Danielka International Space Drug State Corporation of Glen Head, NY
- Danielka International Space Drug State Corporation of Glens Falls, NY
- Danielka International Space Drug State Corporation of Glenwood Landing, NY
- Danielka International Space Drug State Corporation of Gloversville, NY
- Danielka International Space Drug State Corporation of Gorham, NY
- Danielka International Space Drug State Corporation of Goshen, NY
- Danielka International Space Drug State Corporation of Gouverneur, NY
- Danielka International Space Drug State Corporation of Gowanda, NY
- Danielka International Space Drug State Corporation of Grafton, NY
- Danielka International Space Drug State Corporation of Grand Island, NY
- Danielka International Space Drug State Corporation of Granville, NY
- Danielka International Space Drug State Corporation of Great Bend, NY
- Danielka International Space Drug State Corporation of Great Neck, NY
- Danielka International Space Drug State Corporation of Great River, NY
- Danielka International Space Drug State Corporation of Great Valley, NY
- Danielka International Space Drug State Corporation of Greene, NY
- Danielka International Space Drug State Corporation of Greenfield Center, NY
- Danielka International Space Drug State Corporation of Greenlawn, NY
- Danielka International Space Drug State Corporation of Greenport, NY
- Danielka International Space Drug State Corporation of Greenvale, NY
- Danielka International Space Drug State Corporation of Greenville, NY
- Danielka International Space Drug State Corporation of Greenwich, NY
- Danielka International Space Drug State Corporation of Greenwood Lake, NY
- Danielka International Space Drug State Corporation of Greenwood, NY
- Danielka International Space Drug State Corporation of Greig, NY
- Danielka International Space Drug State Corporation of Groton, NY
- Danielka International Space Drug State Corporation of Groveland, NY
- Danielka International Space Drug State Corporation of Guilderland, NY
- Danielka International Space Drug State Corporation of Guilford, NY

- Danielka International Space Drug State Corporation of Hadley, NY
- Danielka International Space Drug State Corporation of Hagaman, NY
- Danielka International Space Drug State Corporation of Hague, NY
- Danielka International Space Drug State Corporation of Hamburg, NY
- Danielka International Space Drug State Corporation of Hamden, NY
- Danielka International Space Drug State Corporation of Hamilton, NY
- Danielka International Space Drug State Corporation of Hamlin, NY
- Danielka International Space Drug State Corporation of Hammond, NY
- Danielka International Space Drug State Corporation of Hammondsport, NY
- Danielka International Space Drug State Corporation of Hampton Bays, NY
- Danielka International Space Drug State Corporation of Hampton, NY
- Danielka International Space Drug State Corporation of Hancock, NY
- Danielka International Space Drug State Corporation of Hannibal, NY
- Danielka International Space Drug State Corporation of Harford, NY
- Danielka International Space Drug State Corporation of Harpersfield, NY
- Danielka International Space Drug State Corporation of Harriman, NY
- Danielka International Space Drug State Corporation of Harris, NY
- Danielka International Space Drug State Corporation of Harrison, NY
- Danielka International Space Drug State Corporation of Harrisville, NY
- Danielka International Space Drug State Corporation of Hartford, NY
- Danielka International Space Drug State Corporation of Hartsdale, NY
- Danielka International Space Drug State Corporation of Hartwick, NY
- Danielka International Space Drug State Corporation of Hastings On Hudson, NY
- Danielka International Space Drug State Corporation of Hastings, NY
- Danielka International Space Drug State Corporation of Hauppauge, NY
- Danielka International Space Drug State Corporation of Haverstraw, NY
- Danielka International Space Drug State Corporation of Hawthorne, NY
- Danielka International Space Drug State Corporation of Hector, NY
- Danielka International Space Drug State Corporation of Hempstead, NY
- Danielka International Space Drug State Corporation of Henderson, NY
- Danielka International Space Drug State Corporation of Henrietta, NY
- Danielka International Space Drug State Corporation of Herkimer, NY
- Danielka International Space Drug State Corporation of Hermon, NY

- Danielka International Space Drug State Corporation of Heuvelton, NY
- Danielka International Space Drug State Corporation of Hewlett, NY
- Danielka International Space Drug State Corporation of Hicksville, NY
- Danielka International Space Drug State Corporation of High Falls, NY
- Danielka International Space Drug State Corporation of Highland Falls, NY
- Danielka International Space Drug State Corporation of Highland Mills, NY
- Danielka International Space Drug State Corporation of Highland, NY
- Danielka International Space Drug State Corporation of Hillburn, NY
- Danielka International Space Drug State Corporation of Hillsdale, NY
- Danielka International Space Drug State Corporation of Hilton, NY
- Danielka International Space Drug State Corporation of Hinsdale, NY
- Danielka International Space Drug State Corporation of Hobart, NY
- Danielka International Space Drug State Corporation of Holbrook, NY
- Danielka International Space Drug State Corporation of Holland Patent, NY
- Danielka International Space Drug State Corporation of Holland, NY
- Danielka International Space Drug State Corporation of Holley, NY
- Danielka International Space Drug State Corporation of Holtsville, NY
- Danielka International Space Drug State Corporation of Homer, NY
- Danielka International Space Drug State Corporation of Honeoye Falls, NY
- Danielka International Space Drug State Corporation of Hoosick Falls, NY
- Danielka International Space Drug State Corporation of Hoosick, NY
- Danielka International Space Drug State Corporation of Hopewell Junction, NY
- Danielka International Space Drug State Corporation of Hornell, NY
- Danielka International Space Drug State Corporation of Horseheads, NY
- Danielka International Space Drug State Corporation of Houghton, NY
- Danielka International Space Drug State Corporation of Howard Beach, NY
- Danielka International Space Drug State Corporation of Hudson Falls, NY
- Danielka International Space Drug State Corporation of Hudson, NY
- Danielka International Space Drug State Corporation of Hume, NY
- Danielka International Space Drug State Corporation of Hunter, NY
- Danielka International Space Drug State Corporation of Huntington Station, NY
- Danielka International Space Drug State Corporation of Huntington, NY
- Danielka International Space Drug State Corporation of Hurley, NY

- Danielka International Space Drug State Corporation of Hyde Park, NY
- Danielka International Space Drug State Corporation of Ilion, NY
- Danielka International Space Drug State Corporation of Indian Lake, NY
- Danielka International Space Drug State Corporation of Inlet, NY
- Danielka International Space Drug State Corporation of Interlaken, NY
- Danielka International Space Drug State Corporation of Inwood, NY
- Danielka International Space Drug State Corporation of Irvington, NY
- Danielka International Space Drug State Corporation of Island Park, NY
- Danielka International Space Drug State Corporation of Islandia, NY
- Danielka International Space Drug State Corporation of Islip Terrace, NY
- Danielka International Space Drug State Corporation of Islip, NY
- Danielka International Space Drug State Corporation of Ithaca, NY
- Danielka International Space Drug State Corporation of Jackson Heights, NY
- Danielka International Space Drug State Corporation of Jamesport, NY
- Danielka International Space Drug State Corporation of Jamestown, NY
- Danielka International Space Drug State Corporation of Jasper, NY
- Danielka International Space Drug State Corporation of Java Center, NY
- Danielka International Space Drug State Corporation of Jay, NY
- Danielka International Space Drug State Corporation of Jefferson Valley, NY
- Danielka International Space Drug State Corporation of Jefferson, NY
- Danielka International Space Drug State Corporation of Jeffersonville, NY
- Danielka International Space Drug State Corporation of Jericho, NY
- Danielka International Space Drug State Corporation of Jewett, NY
- Danielka International Space Drug State Corporation of Johnsburg, NY
- Danielka International Space Drug State Corporation of Johnson City, NY
- Danielka International Space Drug State Corporation of Johnstown, NY
- Danielka International Space Drug State Corporation of Jordan, NY
- Danielka International Space Drug State Corporation of Keene, NY
- Danielka International Space Drug State Corporation of Keeseville, NY
- Danielka International Space Drug State Corporation of Kendall, NY
- Danielka International Space Drug State Corporation of Kent, NY
- Danielka International Space Drug State Corporation of Kerhonkson, NY
- Danielka International Space Drug State Corporation of Kinderhook, NY

- Danielka International Space Drug State Corporation of Kings Park, NY
- Danielka International Space Drug State Corporation of Kingston, NY
- Danielka International Space Drug State Corporation of Kirkwood, NY
- Danielka International Space Drug State Corporation of Knox, NY
- Danielka International Space Drug State Corporation of La Fargeville, NY
- Danielka International Space Drug State Corporation of La Fayette, NY
- Danielka International Space Drug State Corporation of Lacona, NY
- Danielka International Space Drug State Corporation of Lake George, NY
- Danielka International Space Drug State Corporation of Lake Grove, NY
- Danielka International Space Drug State Corporation of Lake Katrine, NY
- Danielka International Space Drug State Corporation of Lake Luzerne, NY
- Danielka International Space Drug State Corporation of Lake Placid, NY
- Danielka International Space Drug State Corporation of Lake Pleasant, NY
- Danielka International Space Drug State Corporation of Lake View, NY
- Danielka International Space Drug State Corporation of Lakewood, NY
- Danielka International Space Drug State Corporation of Lancaster, NY
- Danielka International Space Drug State Corporation of Lansing, NY
- Danielka International Space Drug State Corporation of Larchmont, NY
- Danielka International Space Drug State Corporation of Laurel, NY
- Danielka International Space Drug State Corporation of Laurens, NY
- Danielka International Space Drug State Corporation of Lawrence, NY
- Danielka International Space Drug State Corporation of Le Roy, NY
- Danielka International Space Drug State Corporation of Lee Center, NY
- Danielka International Space Drug State Corporation of Leeds, NY
- Danielka International Space Drug State Corporation of Leicester, NY
- Danielka International Space Drug State Corporation of Leon, NY
- Danielka International Space Drug State Corporation of Levittown, NY
- Danielka International Space Drug State Corporation of Lewis, NY
- Danielka International Space Drug State Corporation of Lewiston, NY
- Danielka International Space Drug State Corporation of Lexington, NY
- Danielka International Space Drug State Corporation of Liberty, NY
- Danielka International Space Drug State Corporation of Lima, NY
- Danielka International Space Drug State Corporation of Limestone, NY

- Danielka International Space Drug State Corporation of Lincolndale, NY
- Danielka International Space Drug State Corporation of Lindenhurst, NY
- Danielka International Space Drug State Corporation of Lindley, NY
- Danielka International Space Drug State Corporation of Lisbon, NY
- Danielka International Space Drug State Corporation of Lisle, NY
- Danielka International Space Drug State Corporation of Little Falls, NY
- Danielka International Space Drug State Corporation of Little Valley, NY
- Danielka International Space Drug State Corporation of Liverpool, NY
- Danielka International Space Drug State Corporation of Livingston Manor, NY
- Danielka International Space Drug State Corporation of Livingston, NY
- Danielka International Space Drug State Corporation of Livonia, NY
- Danielka International Space Drug State Corporation of Locke, NY
- Danielka International Space Drug State Corporation of Lockport, NY
- Danielka International Space Drug State Corporation of Locust Valley, NY
- Danielka International Space Drug State Corporation of Lodi, NY
- Danielka International Space Drug State Corporation of Long Beach, NY
- Danielka International Space Drug State Corporation of Long Lake, NY
- Danielka International Space Drug State Corporation of Lorraine, NY
- Danielka International Space Drug State Corporation of Lowville, NY
- Danielka International Space Drug State Corporation of Lynbrook, NY
- Danielka International Space Drug State Corporation of Lyndonville, NY
- Danielka International Space Drug State Corporation of Lyon Mountain, NY
- Danielka International Space Drug State Corporation of Lyons Falls, NY
- Danielka International Space Drug State Corporation of Lyons, NY
- Danielka International Space Drug State Corporation of Macedon, NY
- Danielka International Space Drug State Corporation of Machias, NY
- Danielka International Space Drug State Corporation of Madison, NY
- Danielka International Space Drug State Corporation of Madrid, NY
- Danielka International Space Drug State Corporation of Mahopac, NY
- Danielka International Space Drug State Corporation of Maine, NY
- Danielka International Space Drug State Corporation of Malden Bridge, NY
- Danielka International Space Drug State Corporation of Malone, NY
- Danielka International Space Drug State Corporation of Malverne, NY

- Danielka International Space Drug State Corporation of Mamaroneck, NY
- Danielka International Space Drug State Corporation of Manchester, NY
- Danielka International Space Drug State Corporation of Manhasset, NY
- Danielka International Space Drug State Corporation of Manlius, NY
- Danielka International Space Drug State Corporation of Mannsville, NY
- Danielka International Space Drug State Corporation of Manorville, NY
- Danielka International Space Drug State Corporation of Marathon, NY
- Danielka International Space Drug State Corporation of Marcellus, NY
- Danielka International Space Drug State Corporation of Marcy, NY
- Danielka International Space Drug State Corporation of Margaretville, NY
- Danielka International Space Drug State Corporation of Marilla, NY
- Danielka International Space Drug State Corporation of Marion, NY
- Danielka International Space Drug State Corporation of Marlboro, NY
- Danielka International Space Drug State Corporation of Martinsburg, NY
- Danielka International Space Drug State Corporation of Maryland, NY
- Danielka International Space Drug State Corporation of Masonville, NY
- Danielka International Space Drug State Corporation of Massapequa Park, NY
- Danielka International Space Drug State Corporation of Massapequa, NY
- Danielka International Space Drug State Corporation of Massena, NY
- Danielka International Space Drug State Corporation of Mastic Beach, NY
- Danielka International Space Drug State Corporation of Mastic, NY
- Danielka International Space Drug State Corporation of Mattituck, NY
- Danielka International Space Drug State Corporation of Maybrook, NY
- Danielka International Space Drug State Corporation of Mayfield, NY
- Danielka International Space Drug State Corporation of Mayville, NY
- Danielka International Space Drug State Corporation of Mc Donough, NY
- Danielka International Space Drug State Corporation of Mc Graw, NY
- Danielka International Space Drug State Corporation of Mechanicville, NY
- Danielka International Space Drug State Corporation of Medford, NY
- Danielka International Space Drug State Corporation of Medina, NY
- Danielka International Space Drug State Corporation of Medusa, NY
- Danielka International Space Drug State Corporation of Melrose, NY
- Danielka International Space Drug State Corporation of Melville, NY

- Danielka International Space Drug State Corporation of Mendon, NY
- Danielka International Space Drug State Corporation of Meridian, NY
- Danielka International Space Drug State Corporation of Merrick, NY
- Danielka International Space Drug State Corporation of Mexico, NY
- Danielka International Space Drug State Corporation of Mid Island, NY
- Danielka International Space Drug State Corporation of Middleburgh, NY
- Danielka International Space Drug State Corporation of Middleport, NY
- Danielka International Space Drug State Corporation of Middlesex, NY
- Danielka International Space Drug State Corporation of Middletown, NY
- Danielka International Space Drug State Corporation of Middleville, NY
- Danielka International Space Drug State Corporation of Milford, NY
- Danielka International Space Drug State Corporation of Mill Neck, NY
- Danielka International Space Drug State Corporation of Millbrook, NY
- Danielka International Space Drug State Corporation of Miller Place, NY
- Danielka International Space Drug State Corporation of Millerton, NY
- Danielka International Space Drug State Corporation of Millport, NY
- Danielka International Space Drug State Corporation of Milton, NY
- Danielka International Space Drug State Corporation of Mineola, NY
- Danielka International Space Drug State Corporation of Minerva, NY
- Danielka International Space Drug State Corporation of Minetto, NY
- Danielka International Space Drug State Corporation of Mineville, NY
- Danielka International Space Drug State Corporation of Minoa, NY
- Danielka International Space Drug State Corporation of Mohawk, NY
- Danielka International Space Drug State Corporation of Moira, NY
- Danielka International Space Drug State Corporation of Monroe, NY
- Danielka International Space Drug State Corporation of Monsey, NY
- Danielka International Space Drug State Corporation of Montauk, NY
- Danielka International Space Drug State Corporation of Montezuma, NY
- Danielka International Space Drug State Corporation of Montgomery, NY
- Danielka International Space Drug State Corporation of Monticello, NY
- Danielka International Space Drug State Corporation of Montour Falls, NY
- Danielka International Space Drug State Corporation of Mooers, NY
- Danielka International Space Drug State Corporation of Moravia, NY

- Danielka International Space Drug State Corporation of Moriah, NY
- Danielka International Space Drug State Corporation of Moriches, NY
- Danielka International Space Drug State Corporation of Morris, NY
- Danielka International Space Drug State Corporation of Morrisonville, NY
- Danielka International Space Drug State Corporation of Morristown, NY
- Danielka International Space Drug State Corporation of Morrisville, NY
- Danielka International Space Drug State Corporation of Mount Kisco, NY
- Danielka International Space Drug State Corporation of Mount Morris, NY
- Danielka International Space Drug State Corporation of Mount Sinai, NY
- Danielka International Space Drug State Corporation of Mount Vernon, NY
- Danielka International Space Drug State Corporation of Munnsville, NY
- Danielka International Space Drug State Corporation of Nanuet, NY
- Danielka International Space Drug State Corporation of Napanoch, NY
- Danielka International Space Drug State Corporation of Naples, NY
- Danielka International Space Drug State Corporation of Narrowsburg, NY
- Danielka International Space Drug State Corporation of Nassau, NY
- Danielka International Space Drug State Corporation of Natural Bridge, NY
- Danielka International Space Drug State Corporation of Nedrow, NY
- Danielka International Space Drug State Corporation of Nelliston, NY
- Danielka International Space Drug State Corporation of Nesconset, NY
- Danielka International Space Drug State Corporation of Neversink, NY
- Danielka International Space Drug State Corporation of New Baltimore, NY
- Danielka International Space Drug State Corporation of New Berlin, NY
- Danielka International Space Drug State Corporation of New City, NY
- Danielka International Space Drug State Corporation of New Hartford, NY
- Danielka International Space Drug State Corporation of New Haven, NY
- Danielka International Space Drug State Corporation of New Hyde Park, NY
- Danielka International Space Drug State Corporation of New Lebanon, NY
- Danielka International Space Drug State Corporation of New Lisbon, NY
- Danielka International Space Drug State Corporation of New Paltz, NY
- Danielka International Space Drug State Corporation of New Rochelle, NY
- Danielka International Space Drug State Corporation of New Suffolk, NY
- Danielka International Space Drug State Corporation of New Windsor, NY

- Danielka International Space Drug State Corporation of New York Mills, NY
- Danielka International Space Drug State Corporation of New York, NY
- Danielka International Space Drug State Corporation of Newark Valley, NY
- Danielka International Space Drug State Corporation of Newark, NY
- Danielka International Space Drug State Corporation of Newburgh, NY
- Danielka International Space Drug State Corporation of Newcomb, NY
- Danielka International Space Drug State Corporation of Newfane, NY
- Danielka International Space Drug State Corporation of Newfield, NY
- Danielka International Space Drug State Corporation of Newport, NY
- Danielka International Space Drug State Corporation of Niagara Falls, NY
- Danielka International Space Drug State Corporation of Nichols, NY
- Danielka International Space Drug State Corporation of Niverville, NY
- Danielka International Space Drug State Corporation of Norfolk, NY
- Danielka International Space Drug State Corporation of North Babylon, NY
- Danielka International Space Drug State Corporation of North Bay, NY
- Danielka International Space Drug State Corporation of North Boston, NY
- Danielka International Space Drug State Corporation of North Collins, NY
- Danielka International Space Drug State Corporation of North Hudson, NY
- Danielka International Space Drug State Corporation of North Norwich, NY
- Danielka International Space Drug State Corporation of North Salem, NY
- Danielka International Space Drug State Corporation of North Tonawanda, NY
- Danielka International Space Drug State Corporation of Northport, NY
- Danielka International Space Drug State Corporation of Northville, NY
- Danielka International Space Drug State Corporation of Norwich, NY
- Danielka International Space Drug State Corporation of Norwood, NY
- Danielka International Space Drug State Corporation of Nunda, NY
- Danielka International Space Drug State Corporation of Nyack, NY
- Danielka International Space Drug State Corporation of Oakdale, NY
- Danielka International Space Drug State Corporation of Oakfield, NY
- Danielka International Space Drug State Corporation of Oceanside, NY
- Danielka International Space Drug State Corporation of Odessa, NY
- Danielka International Space Drug State Corporation of Ogdensburg, NY
- Danielka International Space Drug State Corporation of Olcott, NY

- Danielka International Space Drug State Corporation of Old Bethpage, NY
- Danielka International Space Drug State Corporation of Old Westbury, NY
- Danielka International Space Drug State Corporation of Olean, NY
- Danielka International Space Drug State Corporation of Oneida, NY
- Danielka International Space Drug State Corporation of Oneonta, NY
- Danielka International Space Drug State Corporation of Ontario, NY
- Danielka International Space Drug State Corporation of Orangeburg, NY
- Danielka International Space Drug State Corporation of Orchard Park, NY
- Danielka International Space Drug State Corporation of Orient, NY
- Danielka International Space Drug State Corporation of Oriskany Falls, NY
- Danielka International Space Drug State Corporation of Oriskany, NY
- Danielka International Space Drug State Corporation of Orwell, NY
- Danielka International Space Drug State Corporation of Ossining, NY
- Danielka International Space Drug State Corporation of Oswegatchie, NY
- Danielka International Space Drug State Corporation of Oswego, NY
- Danielka International Space Drug State Corporation of Otego, NY
- Danielka International Space Drug State Corporation of Otisville, NY
- Danielka International Space Drug State Corporation of Otto, NY
- Danielka International Space Drug State Corporation of Ovid, NY
- Danielka International Space Drug State Corporation of Owego, NY
- Danielka International Space Drug State Corporation of Oxford, NY
- Danielka International Space Drug State Corporation of Oyster Bay, NY
- Danielka International Space Drug State Corporation of Painted Post, NY
- Danielka International Space Drug State Corporation of Palatine Bridge, NY
- Danielka International Space Drug State Corporation of Palenville, NY
- Danielka International Space Drug State Corporation of Palmyra, NY
- Danielka International Space Drug State Corporation of Panama, NY
- Danielka International Space Drug State Corporation of Parish, NY
- Danielka International Space Drug State Corporation of Parishville, NY
- Danielka International Space Drug State Corporation of Patchogue, NY
- Danielka International Space Drug State Corporation of Patterson, NY
- Danielka International Space Drug State Corporation of Pavilion, NY
- Danielka International Space Drug State Corporation of Pearl River, NY

- Danielka International Space Drug State Corporation of Peconic, NY
- Danielka International Space Drug State Corporation of Peekskill, NY
- Danielka International Space Drug State Corporation of Pelham, NY
- Danielka International Space Drug State Corporation of Penfield, NY
- Danielka International Space Drug State Corporation of Penn Yan, NY
- Danielka International Space Drug State Corporation of Perry, NY
- Danielka International Space Drug State Corporation of Perrysburg, NY
- Danielka International Space Drug State Corporation of Peru, NY
- Danielka International Space Drug State Corporation of Petersburg, NY
- Danielka International Space Drug State Corporation of Phelps, NY
- Danielka International Space Drug State Corporation of Philadelphia, NY
- Danielka International Space Drug State Corporation of Philmont, NY
- Danielka International Space Drug State Corporation of Phoenicia, NY
- Danielka International Space Drug State Corporation of Phoenix, NY
- Danielka International Space Drug State Corporation of Piercefield, NY
- Danielka International Space Drug State Corporation of Piermont, NY
- Danielka International Space Drug State Corporation of Pike, NY
- Danielka International Space Drug State Corporation of Pine Bush, NY
- Danielka International Space Drug State Corporation of Pine Hill, NY
- Danielka International Space Drug State Corporation of Pine Plains, NY
- Danielka International Space Drug State Corporation of Pitcher, NY
- Danielka International Space Drug State Corporation of Pittsford, NY
- Danielka International Space Drug State Corporation of Plainview, NY
- Danielka International Space Drug State Corporation of Plattekill, NY
- Danielka International Space Drug State Corporation of Plattsburgh, NY
- Danielka International Space Drug State Corporation of Pleasant Valley, NY
- Danielka International Space Drug State Corporation of Pleasantville, NY
- Danielka International Space Drug State Corporation of Plymouth, NY
- Danielka International Space Drug State Corporation of Poestenkill, NY
- Danielka International Space Drug State Corporation of Point Lookout, NY
- Danielka International Space Drug State Corporation of Poland, NY
- Danielka International Space Drug State Corporation of Pomona, NY
- Danielka International Space Drug State Corporation of Pompey, NY

- Danielka International Space Drug State Corporation of Port Byron, NY
- Danielka International Space Drug State Corporation of Port Chester, NY
- Danielka International Space Drug State Corporation of Port Ewen, NY
- Danielka International Space Drug State Corporation of Port Henry, NY
- Danielka International Space Drug State Corporation of Port Jefferson Station, NY
- Danielka International Space Drug State Corporation of Port Jefferson, NY
- Danielka International Space Drug State Corporation of Port Jervis, NY
- Danielka International Space Drug State Corporation of Port Leyden, NY
- Danielka International Space Drug State Corporation of Port Washington, NY
- Danielka International Space Drug State Corporation of Portageville, NY
- Danielka International Space Drug State Corporation of Porter Corners, NY
- Danielka International Space Drug State Corporation of Portland, NY
- Danielka International Space Drug State Corporation of Portville, NY
- Danielka International Space Drug State Corporation of Potsdam, NY
- Danielka International Space Drug State Corporation of Poughkeepsie, NY
- Danielka International Space Drug State Corporation of Pound Ridge, NY
- Danielka International Space Drug State Corporation of Prattsburgh, NY
- Danielka International Space Drug State Corporation of Prattsville, NY
- Danielka International Space Drug State Corporation of Preble, NY
- Danielka International Space Drug State Corporation of Preston Hollow, NY
- Danielka International Space Drug State Corporation of Prospect, NY
- Danielka International Space Drug State Corporation of Pulaski, NY
- Danielka International Space Drug State Corporation of Pulteney, NY
- Danielka International Space Drug State Corporation of Putnam Station, NY
- Danielka International Space Drug State Corporation of Putnam Valley, NY
- Danielka International Space Drug State Corporation of Quogue, NY
- Danielka International Space Drug State Corporation of Randolph, NY
- Danielka International Space Drug State Corporation of Ransomville, NY
- Danielka International Space Drug State Corporation of Ravena, NY
- Danielka International Space Drug State Corporation of Reading Center, NY
- Danielka International Space Drug State Corporation of Red Creek, NY
- Danielka International Space Drug State Corporation of Red Hook, NY
- Danielka International Space Drug State Corporation of Redfield, NY

- Danielka International Space Drug State Corporation of Redford, NY
- Danielka International Space Drug State Corporation of Redwood, NY
- Danielka International Space Drug State Corporation of Remsen, NY
- Danielka International Space Drug State Corporation of Remsenburg, NY
- Danielka International Space Drug State Corporation of Rensselaer Falls, NY
- Danielka International Space Drug State Corporation of Rensselaer, NY
- Danielka International Space Drug State Corporation of Rensselaerville, NY
- Danielka International Space Drug State Corporation of Rhinebeck, NY
- Danielka International Space Drug State Corporation of Richburg, NY
- Danielka International Space Drug State Corporation of Richfield Springs, NY
- Danielka International Space Drug State Corporation of Richford, NY
- Danielka International Space Drug State Corporation of Richland, NY
- Danielka International Space Drug State Corporation of Richmond Hill, NY
- Danielka International Space Drug State Corporation of Richmondville, NY
- Danielka International Space Drug State Corporation of Richville, NY
- Danielka International Space Drug State Corporation of Ridge, NY
- Danielka International Space Drug State Corporation of Rifton, NY
- Danielka International Space Drug State Corporation of Ripley, NY
- Danielka International Space Drug State Corporation of Riverhead, NY
- Danielka International Space Drug State Corporation of Rochester, NY
- Danielka International Space Drug State Corporation of Rock Hill, NY
- Danielka International Space Drug State Corporation of Rockville Centre, NY
- Danielka International Space Drug State Corporation of Rocky Point, NY
- Danielka International Space Drug State Corporation of Rodman, NY
- Danielka International Space Drug State Corporation of Rome, NY
- Danielka International Space Drug State Corporation of Romulus, NY
- Danielka International Space Drug State Corporation of Ronkonkoma, NY
- Danielka International Space Drug State Corporation of Roosevelt, NY
- Danielka International Space Drug State Corporation of Roscoe, NY
- Danielka International Space Drug State Corporation of Rose, NY
- Danielka International Space Drug State Corporation of Roseboom, NY
- Danielka International Space Drug State Corporation of Rosendale, NY
- Danielka International Space Drug State Corporation of Roslyn Heights, NY

- Danielka International Space Drug State Corporation of Roslyn, NY
- Danielka International Space Drug State Corporation of Rotterdam Junction, NY
- Danielka International Space Drug State Corporation of Round Lake, NY
- Danielka International Space Drug State Corporation of Rouses Point, NY
- Danielka International Space Drug State Corporation of Roxbury, NY
- Danielka International Space Drug State Corporation of Rush, NY
- Danielka International Space Drug State Corporation of Rushford, NY
- Danielka International Space Drug State Corporation of Rushville, NY
- Danielka International Space Drug State Corporation of Russell, NY
- Danielka International Space Drug State Corporation of Rye, NY
- Danielka International Space Drug State Corporation of Sackets Harbor, NY
- Danielka International Space Drug State Corporation of Sag Harbor, NY
- Danielka International Space Drug State Corporation of Sagaponack, NY
- Danielka International Space Drug State Corporation of Saint Bonaventure, NY
- Danielka International Space Drug State Corporation of Saint James, NY
- Danielka International Space Drug State Corporation of Saint Johnsville, NY
- Danielka International Space Drug State Corporation of Salamanca, NY
- Danielka International Space Drug State Corporation of Salem, NY
- Danielka International Space Drug State Corporation of Salisbury Center, NY
- Danielka International Space Drug State Corporation of Sand Lake, NY
- Danielka International Space Drug State Corporation of Sandy Creek, NY
- Danielka International Space Drug State Corporation of Sangerfield, NY
- Danielka International Space Drug State Corporation of Saranac Lake, NY
- Danielka International Space Drug State Corporation of Saranac, NY
- Danielka International Space Drug State Corporation of Saratoga Springs, NY
- Danielka International Space Drug State Corporation of Sardinia, NY
- Danielka International Space Drug State Corporation of Saugerties, NY
- Danielka International Space Drug State Corporation of Savannah, NY
- Danielka International Space Drug State Corporation of Savona, NY
- Danielka International Space Drug State Corporation of Sayville, NY
- Danielka International Space Drug State Corporation of Scarsdale, NY
- Danielka International Space Drug State Corporation of Schaghticoke, NY
- Danielka International Space Drug State Corporation of Schenectady, NY

- Danielka International Space Drug State Corporation of Schodack Landing, NY
- Danielka International Space Drug State Corporation of Schoharie, NY
- Danielka International Space Drug State Corporation of Schroon Lake, NY
- Danielka International Space Drug State Corporation of Schuyler Falls, NY
- Danielka International Space Drug State Corporation of Schuyler Lake, NY
- Danielka International Space Drug State Corporation of Schuylerville, NY
- Danielka International Space Drug State Corporation of Scio, NY
- Danielka International Space Drug State Corporation of Scipio Center, NY
- Danielka International Space Drug State Corporation of Scottsville, NY
- Danielka International Space Drug State Corporation of Sea Cliff, NY
- Danielka International Space Drug State Corporation of Seaford, NY
- Danielka International Space Drug State Corporation of Selden, NY
- Danielka International Space Drug State Corporation of Seneca Falls, NY
- Danielka International Space Drug State Corporation of Shandaken, NY
- Danielka International Space Drug State Corporation of Sharon Springs, NY
- Danielka International Space Drug State Corporation of Shelter Island Heights, NY
- Danielka International Space Drug State Corporation of Shelter Island, NY
- Danielka International Space Drug State Corporation of Shenorock, NY
- Danielka International Space Drug State Corporation of Sherburne, NY
- Danielka International Space Drug State Corporation of Sheridan, NY
- Danielka International Space Drug State Corporation of Sherman, NY
- Danielka International Space Drug State Corporation of Sherrill, NY
- Danielka International Space Drug State Corporation of Shirley, NY
- Danielka International Space Drug State Corporation of Shokan, NY
- Danielka International Space Drug State Corporation of Shoreham, NY
- Danielka International Space Drug State Corporation of Shortsville, NY
- Danielka International Space Drug State Corporation of Shrub Oak, NY
- Danielka International Space Drug State Corporation of Sidney, NY
- Danielka International Space Drug State Corporation of Silver Creek, NY
- Danielka International Space Drug State Corporation of Silver Springs, NY
- Danielka International Space Drug State Corporation of Sinclairville, NY
- Danielka International Space Drug State Corporation of Skaneateles, NY
- Danielka International Space Drug State Corporation of Sloatsburg, NY

- Danielka International Space Drug State Corporation of Smyrna, NY
- Danielka International Space Drug State Corporation of Sodus Point, NY
- Danielka International Space Drug State Corporation of Sodus, NY
- Danielka International Space Drug State Corporation of Somers, NY
- Danielka International Space Drug State Corporation of Sound Beach, NY
- Danielka International Space Drug State Corporation of South Dayton, NY
- Danielka International Space Drug State Corporation of South Fallsburg, NY
- Danielka International Space Drug State Corporation of South Glens Falls, NY
- Danielka International Space Drug State Corporation of Southampton, NY
- Danielka International Space Drug State Corporation of Southold, NY
- Danielka International Space Drug State Corporation of Speculator, NY
- Danielka International Space Drug State Corporation of Spencer, NY
- Danielka International Space Drug State Corporation of Spencerport, NY
- Danielka International Space Drug State Corporation of Spring Valley, NY
- Danielka International Space Drug State Corporation of Springville, NY
- Danielka International Space Drug State Corporation of Staatsburg, NY
- Danielka International Space Drug State Corporation of Stafford, NY
- Danielka International Space Drug State Corporation of Stamford, NY
- Danielka International Space Drug State Corporation of Stanfordville, NY
- Danielka International Space Drug State Corporation of Star Lake, NY
- Danielka International Space Drug State Corporation of Stephentown, NY
- Danielka International Space Drug State Corporation of Sterling, NY
- Danielka International Space Drug State Corporation of Stillwater, NY
- Danielka International Space Drug State Corporation of Stockton, NY
- Danielka International Space Drug State Corporation of Stone Ridge, NY
- Danielka International Space Drug State Corporation of Stony Brook, NY
- Danielka International Space Drug State Corporation of Stony Creek, NY
- Danielka International Space Drug State Corporation of Stony Point, NY
- Danielka International Space Drug State Corporation of Stratford, NY
- Danielka International Space Drug State Corporation of Stuyvesant, NY
- Danielka International Space Drug State Corporation of Suffern, NY
- Danielka International Space Drug State Corporation of Summit, NY
- Danielka International Space Drug State Corporation of Sylvan Beach, NY

- Danielka International Space Drug State Corporation of Syosset, NY
- Danielka International Space Drug State Corporation of Syracuse, NY
- Danielka International Space Drug State Corporation of Tannersville, NY
- Danielka International Space Drug State Corporation of Tappan, NY
- Danielka International Space Drug State Corporation of Tarrytown, NY
- Danielka International Space Drug State Corporation of Theresa, NY
- Danielka International Space Drug State Corporation of Thiells, NY
- Danielka International Space Drug State Corporation of Thompsonville, NY
- Danielka International Space Drug State Corporation of Thornwood, NY
- Danielka International Space Drug State Corporation of Ticonderoga, NY
- Danielka International Space Drug State Corporation of Tillson, NY
- Danielka International Space Drug State Corporation of Tioga Center, NY
- Danielka International Space Drug State Corporation of Tomkins Cove, NY
- Danielka International Space Drug State Corporation of Tonawanda, NY
- Danielka International Space Drug State Corporation of Tribes Hill, NY
- Danielka International Space Drug State Corporation of Troupsburg, NY
- Danielka International Space Drug State Corporation of Troy, NY
- Danielka International Space Drug State Corporation of Trumansburg, NY
- Danielka International Space Drug State Corporation of Truxton, NY
- Danielka International Space Drug State Corporation of Tuckahoe, NY
- Danielka International Space Drug State Corporation of Tully, NY
- Danielka International Space Drug State Corporation of Tupper Lake, NY
- Danielka International Space Drug State Corporation of Turin, NY
- Danielka International Space Drug State Corporation of Tuxedo Park, NY
- Danielka International Space Drug State Corporation of Tyrone, NY
- Danielka International Space Drug State Corporation of Ulster Park, NY
- Danielka International Space Drug State Corporation of Unadilla, NY
- Danielka International Space Drug State Corporation of Union Hill, NY
- Danielka International Space Drug State Corporation of Union Springs, NY
- Danielka International Space Drug State Corporation of Uniondale, NY
- Danielka International Space Drug State Corporation of Unionville, NY
- Danielka International Space Drug State Corporation of Utica, NY
- Danielka International Space Drug State Corporation of Vails Gate, NY

- Danielka International Space Drug State Corporation of Valatie, NY
- Danielka International Space Drug State Corporation of Valhalla, NY
- Danielka International Space Drug State Corporation of Valley Cottage, NY
- Danielka International Space Drug State Corporation of Valley Falls, NY
- Danielka International Space Drug State Corporation of Valley Stream, NY
- Danielka International Space Drug State Corporation of Van Buren Point, NY
- Danielka International Space Drug State Corporation of Van Etten, NY
- Danielka International Space Drug State Corporation of Vernon, NY
- Danielka International Space Drug State Corporation of Verona, NY
- Danielka International Space Drug State Corporation of Verplanck, NY
- Danielka International Space Drug State Corporation of Vestal, NY
- Danielka International Space Drug State Corporation of Victor, NY
- Danielka International Space Drug State Corporation of Victory Mills, NY
- Danielka International Space Drug State Corporation of Voorheesville, NY
- Danielka International Space Drug State Corporation of Waddington, NY
- Danielka International Space Drug State Corporation of Wading River, NY
- Danielka International Space Drug State Corporation of Wainscott, NY
- Danielka International Space Drug State Corporation of Walden, NY
- Danielka International Space Drug State Corporation of Wales Center, NY
- Danielka International Space Drug State Corporation of Walker Valley, NY
- Danielka International Space Drug State Corporation of Wallkill, NY
- Danielka International Space Drug State Corporation of Walton, NY
- Danielka International Space Drug State Corporation of Walworth, NY
- Danielka International Space Drug State Corporation of Wampsville, NY
- Danielka International Space Drug State Corporation of Wantagh, NY
- Danielka International Space Drug State Corporation of Wappingers Falls, NY
- Danielka International Space Drug State Corporation of Warrensburg, NY
- Danielka International Space Drug State Corporation of Warsaw, NY
- Danielka International Space Drug State Corporation of Warwick, NY
- Danielka International Space Drug State Corporation of Washington Mills, NY
- Danielka International Space Drug State Corporation of Washingtonville, NY
- Danielka International Space Drug State Corporation of Waterford, NY
- Danielka International Space Drug State Corporation of Waterloo, NY

- Danielka International Space Drug State Corporation of Watertown, NY
- Danielka International Space Drug State Corporation of Waterville, NY
- Danielka International Space Drug State Corporation of Watervliet, NY
- Danielka International Space Drug State Corporation of Watkins Glen, NY
- Danielka International Space Drug State Corporation of Waverly, NY
- Danielka International Space Drug State Corporation of Wawarsing, NY
- Danielka International Space Drug State Corporation of Wayland, NY
- Danielka International Space Drug State Corporation of Wayne, NY
- Danielka International Space Drug State Corporation of Webster, NY
- Danielka International Space Drug State Corporation of Weedsport, NY
- Danielka International Space Drug State Corporation of Wells, NY
- Danielka International Space Drug State Corporation of Wellsburg, NY
- Danielka International Space Drug State Corporation of Wellsville, NY
- Danielka International Space Drug State Corporation of West Babylon, NY
- Danielka International Space Drug State Corporation of West Bloomfield, NY
- Danielka International Space Drug State Corporation of West Haverstraw, NY
- Danielka International Space Drug State Corporation of West Hempstead, NY
- Danielka International Space Drug State Corporation of West Hurley, NY
- Danielka International Space Drug State Corporation of West Islip, NY
- Danielka International Space Drug State Corporation of West Monroe, NY
- Danielka International Space Drug State Corporation of West Nyack, NY
- Danielka International Space Drug State Corporation of West Point, NY
- Danielka International Space Drug State Corporation of West Sand Lake, NY
- Danielka International Space Drug State Corporation of West Sayville, NY
- Danielka International Space Drug State Corporation of West Winfield, NY
- Danielka International Space Drug State Corporation of Westbury, NY
- Danielka International Space Drug State Corporation of Westerlo, NY
- Danielka International Space Drug State Corporation of Westernville, NY
- Danielka International Space Drug State Corporation of Westfield, NY
- Danielka International Space Drug State Corporation of Westford, NY
- Danielka International Space Drug State Corporation of Westhampton Beach, NY
- Danielka International Space Drug State Corporation of Westhampton, NY
- Danielka International Space Drug State Corporation of Westmoreland, NY

- Danielka International Space Drug State Corporation of Westons Mills, NY
- Danielka International Space Drug State Corporation of Westport, NY
- Danielka International Space Drug State Corporation of White Plains, NY
- Danielka International Space Drug State Corporation of Whitehall, NY
- Danielka International Space Drug State Corporation of Whitesboro, NY
- Danielka International Space Drug State Corporation of Whitney Point, NY
- Danielka International Space Drug State Corporation of Willet, NY
- Danielka International Space Drug State Corporation of Williamson, NY
- Danielka International Space Drug State Corporation of Williamstown, NY
- Danielka International Space Drug State Corporation of Williston Park, NY
- Danielka International Space Drug State Corporation of Willsboro, NY
- Danielka International Space Drug State Corporation of Wilmington, NY
- Danielka International Space Drug State Corporation of Wilson, NY
- Danielka International Space Drug State Corporation of Windham, NY
- Danielka International Space Drug State Corporation of Windsor, NY
- Danielka International Space Drug State Corporation of Wolcott, NY
- Danielka International Space Drug State Corporation of Woodbury, NY
- Danielka International Space Drug State Corporation of Woodhull, NY
- Danielka International Space Drug State Corporation of Woodmere, NY
- Danielka International Space Drug State Corporation of Woodridge, NY
- Danielka International Space Drug State Corporation of Woodstock, NY
- Danielka International Space Drug State Corporation of Worcester, NY
- Danielka International Space Drug State Corporation of Wurtsboro, NY
- Danielka International Space Drug State Corporation of Wyandanch, NY
- Danielka International Space Drug State Corporation of Wynantskill, NY
- Danielka International Space Drug State Corporation of Wyoming, NY
- Danielka International Space Drug State Corporation of Yaphank, NY
- Danielka International Space Drug State Corporation of Yonkers, NY
- Danielka International Space Drug State Corporation of York, NY
- Danielka International Space Drug State Corporation of Yorkshire, NY
- Danielka International Space Drug State Corporation of Yorktown Heights, NY
- Danielka International Space Drug State Corporation of Yorkville, NY
- Danielka International Space Drug State Corporation of Youngstown, NY

There are a total of 994 cities in New York. By selling containers to each city through this corporation, the projected profit sales amount to $59.6 trillion annually. To fulfill the New York market, we need 63 International Space State Corporations.

- People might think it's unconventional, but...
- We are the people who live in the real world and see it happening, even if it's illegal.
- We are the people who believe that directing this money to the proper use can bring free healthcare, education, and other benefits to the people.
- We are the people who believe that, through this, we can contribute to space research and colonization of planets.
- We are the people who believe in the prosperity of countries and the world.

To develop and present this project to the public, we need:

1. Financing
2. Selection of professional employees, specialists
3. Selection and purchase of special materials, technologies for development
4. Open government facilities for public access

Revenue from the sales will be directed towards healthcare, military agency expenses, and the space project program created by Georgiy S Garbuz. This program aims to build a national Garbuz Space Academy and national Garbuz Space Academy schools worldwide to grant foster care and other kids access to education. Funding for grants to educate these kids will be derived from directed taxes on Georgiy S Garbuz's science works and all projects.

The project was initially conceived by Georgiy Sergeyevich Garbuz in 2007 and is being developed under the name "Project

For The Future" (G.S. Garbuz) at the National Garbuz Space Academy Trust Corporation

AFTERWORD

I hope by reading this book, you have gained into the comprehensive plan designed to create a better life and future for people worldwide, for all countries, for children, and for the workforce.

In this space revolution, some have made sacrifices, but these sacrifices are destined to be rewarded, bringing about a better future for all. Today, we've decided to unveil the entire project because we believe people globally will start contributing to science, projects, and technology. Our vision is that within the next 10 years, we'll witness the creation of:

- Over 100 thousand international intergalactic space corporations
- Over 500 thousand international intergalactic space state corporations
- Over 100 million city space stores

These entities will form the backbone of the Space Federation, contributing to peace and prosperity worldwide. By purchasing this book, you also contribute to our research and development efforts, making it possible for every child to receive education in space school academies, fostering a healthier future for everyone.

We extend our heartfelt thanks to you. Remember, *"Live, make, and enjoy"* is our motto. *"Za Detey - For Kids"* is what my country and I offer to every nation in the world, inviting you to become allies for the future of humanity.

Stay connected with America, Ukraine, Moldova, Kazakhstan, and Russia. Stay for *"Lyubov"* - Love: love for the country, love for the people, love for the family, and most importantly, love for the kids. This love will pave the way for a prosperous future for every country, every individual, and every child.

Together, we can build a better and more prosperous future for our world and the generations to come. Remember, children are the future leaders who will shape the world.

Sincerely,

Georgiy Sergeyevich Garbuz

Author

www.ingramcontent.com/pod-product-compliance
Lightning Source LLC
Chambersburg PA
CBHW071552210326
41597CB00019B/3214